Sampling Protocol for Monitoring Abiotic and Biotic Characteristics of Mountain Ponds and Lakes

By Robert L. Hoffman, Torrey J. Tyler, Gary L. Larson, Michael J. Adams, Wendy Wente, and Stephanie Galvan

Chapter 2 of
Book 2, Collection of Environmental Data,
Section A, Biological Science

Prepared in cooperation with the:
North Coast and Cascades Network, National Park Service

Techniques and Methods 2-A2

U.S. Department of the Interior
U.S. Geological Survey

U.S. Department of the Interior
Gale A. Norton, Secretary

U.S. Geological Survey
Charles G. Groat, Director

U.S. Geological Survey, Reston, Virginia: 2005

For sale by U.S. Geological Survey, Information Services
Box 25286, Denver Federal Center
Denver, CO 80225

For more information about the USGS and its products:
Telephone: 1-888-ASK-USGS
World Wide Web: http://www.usgs.gov/

Suggested citation:
Hoffman, R.L., Tyler, T.J., Larson, G.L., Adams, M.J., Wente, Wendy, and Galvan, Stephanie, 2005, Sampling protocol for monitoring abiotic and biotic characteristics of mountain ponds and lakes: U.S. Geological Survey Techniques and Methods 2-A2, 90 p.

Contents

Figures

Tables

Conversion Factors

Multiply	By	To obtain
acre	4,047	square meter
acre	0.4047	hectare
foot (ft)	0.3048	meter
gallon (gal)	3.785	liter
gallon (gal)	0.003785	cubic meter
inch (in.)	2.54	centimeter
inch (in.)	25.4	millimeter
million gallons (Mgal)	3,785	cubic meter
mile (mi)	1.609	kilometer
ounce, fluid (fl. oz)	0.02957	liter
pint (pt)	0.4732	liter
quart (qt)	0.9464	liter
square foot (ft^2)	0.09290	square meter

Temperature in degrees Celsius (°C) may be converted to degrees Fahrenheit (°F) as follows:

$$°F=(1.8×°C)+32$$

Temperature in degrees Fahrenheit (°F) may be converted to degrees Celsius (°C) as follows:

$$°C=(°F-32)/1.8$$

Specific conductance is given in microsiemens per centimeter at 25 degrees Celsius (µS/cm at 25°C).

Concentrations of chemical constituents in water are given either in milligrams per liter (mg/L) or micrograms per liter (µg/L).

This page is intentionally blank.

Sampling Protocol for Monitoring Abiotic and Biotic Characteristics of Mountain Ponds and Lakes

By Robert L. Hoffman[1], Torrey J. Tyler[2], Gary L. Larson[1], Michael J. Adams[1], Wendy Wente[1], and Stephanie Galvan[1]

I. Background and Objectives

Preservation of nature in its original state and human appreciation of wilderness conditions are common themes associated with the designation of wild and protected areas (for example, wilderness areas, national parks, areas of critical environmental concern, etc.) in the United States (Cole and Landres 1996; Louter, 1998). Beyond the protection of natural ecosystems and creation of recreational opportunity, wild areas provide great value for scientific study of natural ecosystems and ecosystem processes (Peine, 1990). Studies of wild area ecosystems provide a basis for management of individual land units and an understanding of the natural world (Parsons and Graber, 1990). Arguably, a very important value of wild area studies is the reference point that these areas provide for assessing human impact on the natural world (Franklin, 1987; Peine, 1990; Cole and Landres, 1996).

Although ecosystems are often thought protected from detrimental activities by designation as wild areas, certain activities within and outside wild area boundaries threaten natural resources and ecosystem processes, particularly in and around aquatic environments (Cole and Landres, 1996). For instance, mountain ponds and lakes in wilderness areas of the western United States have low buffering capacities (Eilers and others, 1989) and are therefore particularly susceptible to effects associated with the deposition of airborne pollutants, global climate change, and land-use practices on adjacent lands (Eilers and others, 1989; Cole and Landres, 1996).

Recreational activities and the plans and practices that manage them also may threaten the pristine conditions of mountain ponds and lakes. Recreational impacts on and near aquatic systems often are more extensive than on adjacent terrestrial areas because ponds, lakes, and streams tend to attract certain types of high impact recreational activities

such as camping and fishing (Cole and Landres, 1996). Enhanced levels of use may have direct negative impacts on aquatic resources from compacted soils and trampled vegetation to the removal or redistribution of vegetation and woody debris associated with pack animal grazing and firewood collection and burning (Cole and Landres, 1996). Increased recreational use also can lead to the concentration of human waste near aquatic systems. Equally threatening to aquatic resources are management practices intended to increase recreational opportunities. Cole and Landres (1996) suggested that the greatest disruptions detectable on large spatial scales in wild areas are related to hunting and fishing, and the translocation of fish and wildlife species to improve recreational opportunities. For example, the stocking of fish into mountain lakes of the western United States to increase recreational fisheries has been implicated in reduced abundances and distributions of amphibian populations as well as several amphibian species (Tyler and others, 1998; Knapp and Matthews, 2000; Knapp and others, 2001; Larson and Hoffman 2002), and has affected aquatic invertebrate assemblages (Buktenica and Larson, 1996; Wicklum, 1998; Knapp and others, 2001) and aquatic food-web structure (Wicklum, 1998).

Given the fragile nature of high elevation ecosystems and current concerns about land-use practices, global climate change, and the introduction of exotic and nonnative species (Cole and Landres, 1996), increased efforts to inventory and monitor aquatic resources of the western United States are warranted. In the Pacific Northwest, although the limnological information for many mountain ponds and lakes is limited, current pond and lake data of regional scope typically contain measurements of select physical parameters (for example, surface area and elevation) or records of fish stocking activities. Recognizing that natural resources in protected areas could be valuable indicators of pristine ecosystem condition and health, the National Park Service (NPS), in 1999, initiated a strategy to determine natural resource status and trends on lands administered by NPS. Among the strategy actions considered was the expansion of inventory and monitoring efforts with emphasis on air and water quality monitoring, and included ponds and lakes as important ecosystems to monitor.

[1] U.S. Geological Survey, Forest and Rangeland Ecosystem Science Center, 777 NW 9th Street, Corvallis, Oregon 97330

[2] U.S. Geological Survey, Klamath Field Station, 6935 Washburn Way, Klamath Falls, Oregon 97603

This document describes field techniques and procedures used for sampling mountain ponds and lakes. These techniques and procedures will be used primarily to monitor, as part of long-term programs in National Parks and other protected areas, the abiotic and biotic characteristics of naturally occurring permanent montane lentic systems up to 75 ha in surface area. However, the techniques and procedures described herein also can be used to sample temporary or ephemeral montane lentic sites. Each Standard Operating Procedure (SOP) section addresses a specific component of the limnological investigation, and describes in detail field sampling methods pertaining to parameters to be measured for each component.

II. Sampling Design and Site Selection

[Adapted in part from Summary Report of the workshop entitled "Development of Monitoring Protocols for Mountain Lakes and Ponds within the National Parks," Chairman Stanford Loeb, August 7-9, 2002, Corvallis, Oregon.]

A. Rationale

The number and types of mountain ponds and lakes to sample in a monitoring program as well as how monitoring sites are to be selected (i.e., representative or random) can be problematic. The fact that there are hundreds, possibly thousands, of ponds and lakes in some national parks makes it unrealistic to propose that all of these aquatic systems be sampled. Furthermore, accessibility and the short time that many of these sites are free of ice and snow compound the difficulties associated with effectively monitoring many montane lentic systems. Therefore, the primary issue related to properly monitoring mountain ponds and lakes is how best to achieve the goal of the monitoring program, that is, how best to determine long-term trends in the limnological characteristics of these aquatic resources.

The selection of sites to be sampled for a monitoring program logically would be based on which ecosystem stressors (e.g., global climate change, transport of anthropogenic contaminants via the atmosphere, introduction of exotic species, and recreational use) are being evaluated and on the monitoring questions being asked. In order to assess long-term trends in the physical, chemical, and biological characteristics of lentic ecosystems, a group of representative sites, repeatedly sampled intensively (i.e., monthly during the ice- and snow-free season or at least annually), could provide fine scale temporal resolution of trends. These sites can provide trend information relatively soon (i.e., 3-5 years) after the monitoring program is started. The National Science Foundation Long-Term Ecological Research Program has adopted this approach. An alternative method would be to randomly select the sites to be surveyed. The advantage of this method is that probabilistic statements can be made about differences in means, and provides a wider range of statistical inference than the representative site approach.

The ability to make statistical inferences to a large population of ponds and lakes does, at first, make the random-selection method appear preferable. This method can add strength to a statement concerning any observed change in a population of ponds and lakes. The representative method has little strength statistically to extrapolate to a larger population of lentic systems. The disadvantage of the random-selection method is that in order to make useful probabilistic statements about responses in a larger population of ponds and lakes, the number of sites in the sampling program would need to be relatively large. Also, if randomly selected monitoring sites are sampled on a rotating basis over a number of years, then the statistical power of inference relative to trend will not be realized until after the second complete rotation of the monitoring cycle. For instance, if 50 sites are sampled over a 5-year period at the rate of 10 sites per year, reliable trend information and statistical inference will not be available until after all sites have been sampled at least twice (i.e., 10 years).

Given the difficulties associated with the ability to sample montane ponds and lakes, implementing a sampling design based only on the random selection of monitoring sites may not be entirely acceptable for a monitoring program, nor feasible both logistically and monetarily. However, adoption of a sampling design based on the selection of representative monitoring sites should not rule out the need to incorporate some form of randomized survey of ponds and lakes into the monitoring program. Therefore, a monitoring program that incorporates both representative and randomly selected sites in its sampling design could meet the program objectives of (1) identifying long-term trends and potential changes in the physical, chemical, and biological characteristics of lentic ecosystems; and (2) establishing the statistical power to infer identified trends and potential changes to the population of lentic ecosystems throughout a given area (e.g., a national park).

B. Grouping Sites for Selection into Sampling Design

Ponds and lakes distributed across the landscape vary relative to physical characteristics such as size, shape, maximum depth, elevation, aspect, and basin geology. They can range from small-shallow/large-deep low elevation to small-shallow/large-deep high elevation systems. Ponds and lakes also vary in terms of their chemical and biological characteristics, and variation of these parameters generally can be related to variation in physical characteristics. In this context, we could hypothesize that lentic systems having similar physical characteristics generally will be similar in terms of their chemical and biological characteristics (Larson and others, 1994, 1995, 1999; Hoffman and others, 1996).

The number of lentic systems and the proportion of systems with relatively similar physical characteristics varies among locations (e.g., national parks). Therefore, prior to selecting monitoring sites, the population of ponds and lakes within a defined area (e.g., national park) needs to be organized into ecologically meaningful groups. Because the size and elevation of many lentic systems usually are recorded in park databases, park ponds and lakes can be grouped relative to these physical characteristics. The process begins with the creation of a database, preferably as an Excel file, consisting of perennial ponds and lakes for which surface area and elevation are known. Sites included in the database should not be lower than 700 m in elevation. The ponds and lakes are then sorted by size into four strata: Stratum 1 = <0.2 ha; Stratum 2 = 0.2 to <0.8 ha; Stratum 3 = 0.8 to <4.0 ha; Stratum 4 = ≥4.0 ha. Once this is completed, the five lakes to be non-randomly selected for intensive sampling should be removed from the four size-based strata. The remaining ponds and lakes in the four size-based strata are sorted by elevation. The results create a table of ponds and lakes ordered by ascending elevation within each of the four size-based strata. At this point, random-selection sites can be chosen for inclusion in the monitoring program.

C. Sampling Design Structure and Site Selection

The sampling design has a two-level structure:

Table 1. Sampling Design.

[IS, intensively sampled; RS, randomly selected]

Year	Level 1 IS	Level 2				
		RS1	RS2	RS3	RS4	RS5
1	5	10				
2	5		10			
3	5			10		
4	5				10	
5	5					10
6	5	10				
7	5		10			
8	5			10		
9	5				10	
10	5					10

Level 1 consists of five representative sites to be intensively sampled (IS). These sites should be *non-randomly* selected from the four size-based strata. Selected IS sites should be relatively easy to access by sampling crews and be either (1) generally representative of the mean size and elevation of the sites in the four size-based strata; or (2) previously sampled sites and/or sites that are of interest to the long-term monitoring program resource managers.

Level 2 is composed of 50 sites that are sampled over a period of 5 years (table 1). The 50 sites are *randomly* selected from the four size-based strata. The following process should be used to randomly select the ponds and lakes to be sampled for each stratum:

1. Sort ponds/lakes into the four sized-based strata;

2. Select the five IS sites (table 1, Level 1) and remove them from the four size-based strata;

3. Sort remaining ponds/lakes in the four size-based strata by ascending elevation;

4. Determine the percentage of total ponds/lakes represented by the number of ponds/lakes in each stratum (table 2, column 3);

5. Multiply 50 (the total number of lakes to be randomly selected for monitoring from all strata) by the percentage determined in step 4 above for each stratum (for example: Stratum 1 = 50 ponds/lakes × 53.7 percent = 26.85 or 27 ponds/lakes; Stratum 2 = 50 ponds/lakes × 26.7 percent = 13.35 or 13 ponds/lakes; etc.);

6. Assign site numbers from 1 to N to each pond/lake in each stratum (***Note**: Ponds/lakes are now ordered by increasing elevation*);

7. Divide the total number of sites to be selected from each stratum by the total number of ponds/lakes in the stratum (e.g.: Stratum 1 (table 2) has a total of 161 ponds/lakes and 27 sites are to be selected; so 27 sites divided by 161 ponds/lakes equals 0.1677);

8. **Randomly** select a start number between 0 and 1 and begin selecting the monitoring sites from each stratum (***Note**: A unique start number should be selected for each stratum*).

    A. Continuing with the Stratum 1 example: a value of 0.606, for instance, is **randomly** selected from between 0 and 1 as the **start number** for this stratum;

    B. The first monitoring site is then selected by dividing 0.606 by 0.1677 = 3.6, which corresponds with site number 3 in Stratum 1 (***Note**: Always round down to whole number*);

    C. For successive site selections, add 1 to the **random start number** and then divide by 0.1677 until all sites have been selected, for example
    site 2: 0.606 + 1 divided by 0.1677 = 9.6 [pond/lake 9];
    site 3: 1.606 + 1 divided by 0.1677 = 15.5 [pond/lake 15];
    site 4: 2.606 + 1 divided by 0.1677 = 21.5 [pond/lake 21]......site 27 = 25.606 + 1 divided by 0.1677 = 158.69 [pond/lake 158]).

9. After the 50 sites have been **randomly** selected for the four size-based strata, sites from each stratum are placed into each of five random selection (RS) groups (Level 2, table 1) so that each RS group contains 10 sites.

Table 2.   MORA ponds/lakes by four size-based strata (an example).

[note: column 3 = number of ponds/lakes per stratum divided by TOTAL; column 5 = number of ponds/lakes to choose per stratum divided by number of ponds/lakes per stratum.]

Stratum	Number of ponds/lakes	Percentage of total ponds/ lakes all strata	Number of ponds/lakes to choose	Percentage of total ponds/lakes per stratum
1	161	53.7	27	16.8
2	80	26.7	13	16.3
3	49	16.3	8	16.3
4	10	3.3	2	20.0
Total	300			

The sampling design structure requires that 15 sites be sampled per year. The five Level 1 IS sites are sampled at least annually and can be sampled monthly during the ice- and snow-free season. The 10 sites in each Level 2 RS group are sampled once every 5 years. During the year in which a RS group is sampled, each site is sampled at least once (twice if sufficient funds are available) during the sampling season. One complete Level 2 sampling cycle (i.e., all 50 randomly selected sites have been sampled) is accomplished after the fifth sampling season.

The two-level structure of this sampling design (table 1) makes possible the singular implementation of either level if for some reason full implementation of this sampling design is not possible or desirable (e.g., limited availability of funding or emphasis on specific questions of interest not associated with a particular design-level). Level 1 (table 1) could be implemented if questions associated with trend and the timely detection of trend were of primary importance, whereas Level 2 (table 1) could be implemented if the ability to infer monitoring results park-wide was of primary importance. Implementation of both levels would provide specific and timely trend information as well as the capacity to infer results of the monitoring program park-wide. The decision on how to implement the proposed sampling design (i.e., Level 1 only, Level 2 only, or 2-Level design) should be made at the park-level based on monitoring needs and availability of funding for sampling design implementation.

III. Samples to be Collected and Timing of Site Visits

[Adapted in part from Summary Report of the workshop entitled "Development of Monitoring Protocols for Mountain Lakes and Ponds within the National Parks," Chairman Stanford Loeb, August 7-9, 2002, Corvallis, Oregon.]

The monitoring program has one overall goal: determination of long-term trends in water quality of the lentic resources in national parks. Within this overall goal, many objectives will be determined based on the specific questions that are being posed. These objectives are focused on currently known types of stressors that might have a negative impact on water quality. However, the parameters monitored over the long-term will represent enough pond and lake characteristics for results of sampling to be applicable to potential problems associated with presently unknown stressors. It is important to appreciate the value of regularly-collected data on even one aquatic system as there is very little such data.

Water quality is defined as the physical, chemical, and biological characteristics of water. The monitoring program will need to have a mixture of parameters representing all the characteristics of water quality. Although this initial list of parameters for the monitoring program may be short, additional parameters should be added when specific questions and/or hypotheses are clearly set forth. Ongoing discussions and modifications to the monitoring program are an integral part of an Adaptive Management approach.

At a minimum, the following physical characteristics of water quality should be monitored:
- Water column temperature;
- Secchi disc transparency; and
- Lake water level.

The bathymetry of each site should also be mapped at least once at the beginning of the monitoring program. The following chemical characteristics should be measured:
- Dissolved inorganic nitrogen (nitrate and ammonium);
- Total phosphorus;
- Chlorophyll;
- Dissolved organic carbon;
- Acid neutralizing capacity;
- pH;
- Conductivity;
- Dissolved oxygen; and
- Total dissolved solids.

Biological characteristics that should be measured include:
- Amphibians;
- Zooplankton;
- Aquatic macroinvertebrates;
- Fish;
- Macrophytes; and
- Epilithic periphyton.

The intensity of recreational activity on and around each lake also needs to be quantified.

Other parameters that could be included in the monitoring program and considered in the survey sampling effort include: (1) cations and anions (e.g., calcium, chlorine, magnesium, mercury, potassium, sodium, and sulfate); (2) pesticides (e.g., hexachlorobenzene, hexachlorocyclohexanes, DDT and its metabolites, polychlorobiphenyls, and polycyclic aromatic hydrocarbons); and (3) phytoplankton.

The selection of parameters to be sampled is based on sensitivity and specificity of these measures in relation to potential influence of environmental stressors. The monitoring program should measure parameters that would be sensitive to a given stressor [e.g., the date of ice-out would be sensitive to global warming; or changes in the taxonomic composition of aquatic organisms would be a more sensitive, early indicator of environmental change than would be aggregate functional variables, such as total biomass (see Schindler, 1987)]. Parameters also should be specific to a stressor, that is, not strongly influenced by unrelated conditions (e.g., the lowering of lake water pH and atmospheric acid deposition). These two attributes should be considered when reviewing which parameters should be collected in the monitoring program. For a parameter to be a valid "indicator" of change in water quality, it ideally needs to be both sensitive and specific.

Ideally, all monitoring sites should be visited twice (i.e., once in July and in August) during each sampling season; however, the five Level 1 IS sites can be sampled each month during this period (i.e., additionally in June if accessible and in September). The July sampling period should begin as soon after ice-out as possible and conclude 3-4 weeks after the sampling visits are begun. The August sampling period should begin during the first full week of August and conclude 3-4 weeks later. Attributes to be sampled at each site during each sampling period are identified in Appendix II. The timing of sampling (i.e., time-of-season, time-of-day) can affect the variability associated with the measurement of various water-quality parameters (e.g., water clarity, chlorophyll-a, nutrients). Every attempt should be made to consistently sample monitoring sites according to the time-of-season of the first sampling visit to each site, and collect samples for water-quality parameters as close to the same time-of-day as possible for all monitoring sites and during subsequent sampling visits.

IV. Site Access and Crew Safety

A. Access to Sampling Sites

Management regulations of U.S. National Parks and Wilderness Areas exclude the use of off-road mechanized transportation (e.g., motorcycles and bicycles), so hiking is the primary method used for accessing backcountry montane ponds and lakes. Under certain circumstances (e.g., excessive

gear-weight and time constraints) helicopters may be used to transport crew members to backcountry study sites. This use is subject to permission from national park and wilderness area managers and/or administrators. Crew members involved in projects using helicopters for access to remote backcountry sites will be required to complete Interagency Aviation B-3 level training that includes training modules A101, A105, A106, A108, and A113. Potential obstacles to safe and easy access to backcountry ponds and lakes include: (1) ruggedness of terrain; (2) vagaries of summer weather in montane areas (especially, early and late summer snowstorms and thunderstorms); and (3) time of pond and lake ice-out.

B. Sampling Crew Safety

A two-person crew is usually adequate for safe and efficient sampling of most mountain ponds and lakes. However, a three-person crew is preferable for transporting necessary field equipment and personal gear to backcountry sampling sites, and for increased crew safety. Additional crew members may be needed periodically if the amount or weight of equipment increases due to additional sampling needs or extended periods in the field.

Assembling a sampling crew of safe and responsible individuals is a top priority of the project manager. Sampling crew supervisors should be experienced field technicians or scientists. Crew members should be capable of living and working cooperatively with others under often stressful and challenging conditions in rugged and isolated areas for extended periods. Under these conditions, risk of personal injury increases and crews are often hours or days removed from emergency medical help. Although the sampling crew's primary objective is to gather data at mountain ponds and lakes, individuals are responsible for personal safety and the safety of other crew members while maintaining positive work relationships.

Safety training increases the safety awareness of crew members and prepares them for responding to potential medical emergencies. Each crew member should be trained, skilled, and/or certified in

- Basic first aid and cardiopulmonary resuscitation (CPR)
- Swimming
- Backpacking skills, wilderness survival, and orienteering
- Use of handheld communication devices
- Aircraft safety awareness
- Boat safety, especially as this relates to the use of inflatable boats

Crew members also should be oriented to and trained in the safe and proper use of sampling equipment and chemicals to be used for water-quality analysis or organism preservation.

SOP 1. Physical Characteristics of Monitoring Sites
Version 1.00

Revision History Log:

Previous Version Number	Revision Date	Author	Changes Made	Reason for Change	New Version Number

Recommended references:

Lind, O.T., 1979, Handbook of common methods in limnology (2nd ed.): The C.V. Mosby Co., London.

Wetzel, R.G., and Likens, G.E., 2000, Limnological analyses (3rd ed.): Springer, New York, 429 p.

A. *Location Coordinates* are determined using a Global Positioning System (GPS) unit. The GPS unit should be set to the North American Datum 1927 and coordinates should be recorded in UTM. The UTM zone also should be recorded (*Note: All sites within the North Coast and Cascades Network are in UTM Zone 10*). Time-Frame (TF) = first visit.

B. *Elevation* can be obtained from existing topographic maps or measured using a handheld altimeter or the altimeter function of a handheld GPS unit. If a handheld altimeter is used to measure elevation, the instrument will need to be benchmark calibrated before each use. Unit of measure is meters. TF = first visit.

C. *Surface Area* is measured (estimated) using polar planimetry on 7.5-min and 15-min USGS topographic maps. Polar planimetry involves the use of a planimeter or electronic digitizer. The protocol for polar planimetry and use of a planimeter are described in Wetzel and Likens (2000, p. 11). If polar planimetry cannot be used to estimate the surface area of a site, alternative methods are available. One method is to determine the mean width and maximum length, in meters, of the monitoring site and then calculate an estimated surface area using the formula:

surface area = site mean width divided by site maximum length (Wetzel and Likens, 2000, p. 9-11). Various GPS units also can be used to determine surface area. The unit manual should be used to guide field technicians in the proper use of the instrument for determining surface area. Surface area is expressed in hectares. TF = every 5 years.

D. *Site Perimeter* is determined by measuring the entire shoreline length of a site using a metered tape or meter-marked line. Length measurements should be made at the shoreline water-land interface. Unit of measure is meters. TF = every 5 years.

E. *Site Inlets and Outlets* are identified during the measurement of site perimeter. Location coordinates of inlets and outlets should be determined using a GPS unit and they also can be identified on a representational or triangulated map of the monitoring site (see Lind, 1979 or Wetzel and Likens, 2000, for map triangulation protocol). TF = every 5 years.

F. *Site Basin Aspect* is determined using a compass or GPS unit. Aspect refers to the orientation of the main axis of a site and is often related to prevailing seasonal winds at the site location. Because many mountain ponds and lakes are located in relatively steep-sided, partially or semi-enclosed basins, aspect also is often related to the direction of the basin as it opens out from the semi-enclosed portion of the basin. Unit of measure is compass direction. TF = first visit.

G. *Basin Watershed Area* is derived from electronic digitization of 7.5-min or 15-min USGS topographical maps. Unit of measure is hectares. TF = first visit.

H. *Basin Geologic Composition* generally can be determined using available GIS data layers. However, for more specific determination, samples of basin geologic substrates (e.g., as rocks and rock fragments) can be collected and returned to the laboratory for further analysis and identification. TF = first visit.

I. *Site Basin Origin (i.e., morphogenetic type)* in montane systems include bench, cirque, tarn, ice scour, kettle, moraine, slump, trough, and fault-influenced. *Bench* sites result from glacial action working at two different time periods to create a step or bench along a hillside, and they typically are oblong in shape and relatively shallow. *Cirque* sites occur in amphitheater-shaped depressions at the head of glaciated valleys and typically occupy most of the depression. *Tarn* sites are small and very shallow and are found within cirque basins. *Ice scour* sites are small, irregularly-shaped, and form in fault or joint fracture lines. They typically occur on ridgetops. *Kettle* sites result from melting remnant pieces of ice left in the outwash of retreating glaciers and occur, often, as small water-bodies left in the depression. *Moraine* sites are formed by unconsolidated glacial deposits left at the face of a glacier (terminal moraine) and(or) on the sides of a glacier (lateral moraine); lakes form behind these deposits. *Slump* sites occur in the pocket formed after the mass movement of soil on a hillside. *Trough* sites are formed in depressions occurring in U-shaped glaciated valleys. *Fault-influenced* sites are formed in the depression left by single fault displacement between two lithic (i.e., of, relating to, or made of stone) masses. [See Liss and others, 1995, Ecological effects of stocked trout in naturally fishless high mountain lakes, North Cascades National Park Service Complex, WA, USA: Technical Report NPS/PNROSU/NRTR-95-03. Copies available from Denver Service Center, Technical Information Center, PO Box 25287, Denver, CO 80225-0287 telephone: (303) 969-2130; see also Hutchinson, G.E., 1957, A treatise on limnology, vol. I. New York, John Wiley and Sons, Inc.]. TF = first visit.

J. *Site Vegetation Zone Categories* include: low elevation forest; high elevation forest; subalpine meadow; and alpine. Low and high elevation forest sites are distinguished by the extent of winter snowpack. Sites in high elevation forest have continuous winter snow cover while low elevation forest sites do not. Subalpine typically refers to the area between forest line and scrub line occupied by a mosaic of tree patches and montane meadow communities, but also can include the closed forest subzones usually found just below subalpine parklands (see Franklin and Dyrness, 1973, p. 248-290). Alpine sites are found in meadows dominated by bare rock and rubble. Snowfields and glaciers also occur in this zone (Franklin and Dyrness, 1973). TF = first visit.

K. *Dominant Vegetation* (i.e., trees, shrubs, grasses, herbs) occurring in site basins should be recorded. Vegetation should be identified to species when possible. Initial vegetation information can be obtained from GIS vegetation data layers, however, more specific information can be obtained by gathering samples of vegetation for laboratory analysis, identification, and confirmation. TF = every 5 years.

L. *A Visual Record* of site physical characteristics can be developed by photographing each site and its basin, preferably using a digital camera. The same site location should be used to photograph the whole lake, basin, and shoreline features. Shoreline photographs should be of prominent features (e.g., talus slope, areas of coarse wood, human disturbance) that could potentially change through time. An accurate record of photo number, photo purpose, and subject should be maintained using the appropriate field data sheet. TF = every 5 years.

M. *All Measurements and Observations* should be recorded on appropriate data sheets and in the monitoring project field book.

SOP 2. Bathymetry, Maximum Depth, Water Level
Version 1.00

Revision History Log:

Previous Version Number	Revision Date	Author	Changes Made	Reason for Change	New Version Number

Recommended references and website:

Lind, O.T., 1979, Handbook of common methods in limnology (2nd ed.): The C.V. Mosby Co., London.

Wetzel, R.G., and Likens, G.E., 2000, Limnological analyses (3rd ed.): Springer, New York, 429 p.

Forest Service Global Positioning System available online at www.fs.fed.us/database/gps.

A. Bathymetry

1. Mapping should be completed at least once at the beginning of the monitoring program and then every 5 to 10 years, thereafter.

2. Two crew members working from an inflatable boat create a bathymetric map of a monitoring site by measuring depths along transects that are parallel and perpendicular to the longest main axis of the site.

    a. The ends of each transect should extend from the land-water interface on one shore to the land-water interface of the opposite shore.

    b. A minimum of three parallel and three perpendicular transects should be included in this survey.

    c. Large lakes (i.e., >1 ha) may require additional transects to accurately record contour intervals.

3. Radial transects can be used to make bathymetric measurements for large (i.e., >1 ha), irregularly shaped lakes.

    a. These transects originate from a single point located at the approximate center of the lake and extend shoreward along the 8 cardinal compass directions (N, NE, E, SE, S, SW, W, NW). These 8 transects are sufficient for lakes up to about 1.5 ha in surface area. If the site is larger than about 1.5 ha then depths along up to 8 additional transects should be measured.

    b. The UTM coordinates of the point of origin should be recorded so that this point can be relocated during subsequent sampling visits to the site.

4. In 2 and 3, a handheld sonar device is used to record depths along each transect. A calibrated line also can be used to measure depth, if a handheld sonar device is not available.

    a. At sites <1 ha, depth measurements should be made at approximately 3 m intervals.

    b. At sites ≥1 ha, depth measurements are made at approximately 10 m intervals.

5. At shallow sites, bathymetry and maximum depth measurements can be made using the handheld sonar device or calibrated line during a wading survey of parallel and perpendicular transects within the shoreline-perimeter of the site.

  a. Wading surveys are best conducted after other samples have been collected to prevent interference with the measurement of other physical, chemical, or biological variables.

  b. It also is recommended that wading surveys be used minimally, and that some type of flotation device be used whenever possible.

  c. As for larger, deeper sites, measurement number and its corresponding depth and UTM coordinates are recorded on the appropriate data sheet.

6. During surveys, measurement numbers and their corresponding depths and UTM coordinates are recorded on the appropriate data sheet. Measurement numbers should be simple integers usually starting with 1.

B. Maximum Depth: The handheld sonar device or calibrated line also is used to determine a site's deepest point. When determined, the maximum depth and the UTM coordinates of its location are recorded on the appropriate data sheet.

C. Useful information concerning trip planning for the use of GPS units and trip planning software can be obtained by accessing the USDA Internet site referenced above (USFS GPS page). Once accessed, click on What's New and then on New Mission Planning/Satellite Availability Software.

D. If a GPS unit is not available, crew members can create a bathymetric map of a site using the plane table triangulation method described in Lind (1979, p. 6-11) or Wetzel and Likens (2000, p. 7-9). This method requires the creation of a relatively accurate representation of the site's perimeter, and then performing the activities described in **A**.

E. Site Water-Level Measurement:

1. Water-level measurements should be made at the five Level 1 IS sites (see table 1).

2. Three bench mark locations should be established near the land-water interface at each of the five monitoring sites. Bench marks should not be more than approximately 15 m from this interface. The metal tag signifying a bench mark should be attached to a permanent object (e.g., tree, bedrock outcrop, etc.). Multiple bench marks are set at each site in case one or more of the bench marks is lost.

3. Water level should be measured from each bench mark during the first season that a pond/lake is sampled, and then from only one bench mark in subsequent years. (*Note: In subsequent years, water level should be measured from the same benchmark*). Measurement of water level should be completed at the same time during each sampling season (i.e., if the measurement was made during the first week of August then water level should be measured at this time in subsequent years).

4. Water-level measurements are made using parachute cord, a line-level, and a measuring staff marked in centimeters that is at least 2 m long. The staff can be telescoping.

5. Measuring water-level:

    a. This process will require at least two crew members.

    b. Attach one end of the parachute cord to the bench mark.

    c. Extend the cord from the bench mark to at least 1 m beyond the land-water interface. **This distance should be recorded and will be the extended distance of the cord during all subsequent sampling visits.**

    d. Attach the line-level to the cord and make certain that the cord is level.

    e. At the distal point of the extended cord, measure the distance from the cord to the surface of the water using the measuring staff.

    f. Record this distance in centimeters

    g. Repeat b. through f. two times

6. Computation of water-level:

    a. The height of the bench mark is arbitrarily designated as 100 m.

    b. Transform the recorded water-level measurement from centimeters into meters and subtract from 100 m.

    c. Example: (1) water-level measurement = 73 cm; (2) transformed water-level measurement = 0.73 m; (3) calculated water-level: 100 m – 0.73 m = 99.27 m.

    d. Repeat a. and b. for two additional measurements.

    e. Calculate mean water-level as sum of the three water-level measurements divided by 3.

    f. Record the three water-level measurements (in centimeters), the three calculated water-levels (in meters), and the calculated mean water-level (in meters).

7. Special Considerations:

    a. Always make water-level measurements from the same bench mark.

    b. If a new bench mark has to be used, a new water-level data set should be created using this new bench mark as the standard.

    c. The use of a new bench mark for measuring water level should be clearly noted in the water-level database.

    d. Although all bench marks will be set at 100 m, water levels using different bench marks cannot be compared because each bench mark will most likely be at a different height above the land-water interface when the standard for measuring water level is initially set.

8. An instrument known as a water-level logger is available for the continuous measurement of water-level. This battery powered instrument uses a pressure transducer and data logger to measure water level. The instrument is available through Global Water Instrumentation, Inc. (**Note:** *This information is not provided as an endorsement of this product*).

SOP 3. Water Temperature
Version 1.00

Revision History Log:

Previous Version Number	Revision Date	Author	Changes Made	Reason for Change	New Version Number

A. Profiles are conducted at or near the deepest point of each site.

B. A remote probe (i.e., thermistor) attached to the end of a transmission wire marked at 0.5-m intervals and attached to a handheld adjustable thermocouple is used to measure water temperature. **The thermistor should be calibrated in the laboratory using a precision glass thermometer.** To maintain a vertical line between the descending remote probe and the water surface, a small, rock-filled stuff sack is attached to the probe.

C. Temperature readings are recorded when thermocouple readings remain constant for approximately 3 seconds. Periodic checking of thermistor readings at the water surface during sampling using a precision glass thermometer will help assure thermistor accuracy and stability. Thermistor and glass thermometer measurements should be clearly identified when recorded. If there are substantial discrepancies between the thermistor and the precision glass thermometer temperature measurements, the thermistor can be recalibrated in the laboratory prior to the next study site sampling visit.

D. Temperature should be measured and recorded in Celsius and corresponding depths in meters.

E. Measurements of water temperature and corresponding depths are made from an inflatable boat with two crew members. One crew member maintains the position of the boat, while the other crew member measures and records temperatures and depths. These measurements are recorded on the appropriate data sheet.

F. Additional information to be recorded when measuring water temperature includes: air temperature (in Celsius), cloud cover (estimated percent), and time of day.

G. A float tube may be used instead of an inflatable boat. In this case the crew member in the float tube will measure temperatures and corresponding depths, and verbally transmit these measurements to an onshore crew member, who will record the data.

H. Temperature and corresponding depth should be measured at just below the surface, at mid-depth in the water column, and at just above the bottom of the site.

I. To measure the water temperature of shallow sites, the temperature probe of the digital thermometer is attached to a telescoping pole (Girdner and Larson, 1995). Temperature recordings should be performed as near to the estimated deepest point of each site as possible.

J. At the discretion of the project supervisor, a more discrete temperature profile (i.e., at 1-m intervals from just below surface to just above bottom) can be performed at sites that are *thermally stratified*. A *thermally stratified* site is one in which a warmer upper layer of water (i.e., epilimnion) is separated from a lower colder layer of water (i.e., hypolimnion) by a boundary where temperature changes at a rate of 1°C or greater per meter (i.e., metalimnion).

K. The metal sensor of the thermistor should be cleaned with 95 percent ethanol before being used at a new monitoring site.

L. At the discretion of the project supervisor, permanently placed mini-thermistors can be used to measure temperature at specific sites. The mini-thermistors should be placed at the depths identified in section H above, or at the depth intervals chosen for the measurement of water column temperature in *thermally stratified* sites (see section J above). Mini-thermistor data logs should be collected annually.

SOP 4. Water Clarity
Version 1.00

Revision History Log:

Previous Version Number	Revision Date	Author	Changes Made	Reason for Change	New Version Number

A. Measure water clarity using a weighted 20 cm black and white Secchi disk that is lowered through the water column at the deepest point of a site.

B. The measurement should be taken from the sunny side of the rubber raft or flotation device to avoid interference from the underwater shadow of the raft or device.

C. The Secchi disk is attached to the end of a nylon or cotton rope marked at 0.5-m intervals. Just below the Secchi disk and also attached to the rope is a weight or small stuff sack filled with rocks that provides weight to the rope as the Secchi disk is lowered through the water column.

D. Secchi disk depth is the average of the recorded descending reading at the point where the disk disappears from the view of an observer and the ascending reading at the point when the disk reappears to the view of the observer. Three of these average measurements should be determined for a site during each visit and recorded on the appropriate data sheet. Data to be recorded for each of the three measurement trials include descending depth, ascending depth, and average depth [i.e., (descending depth + ascending depth)/2 = average depth].

E. Secchi disk depth measurements should be taken when the surface of the site is calm and as close to noon (standard time) as possible. However, if this is not possible, Secchi disk depth measurement can be completed before or after noon under less than calm surface conditions. The time and surface condition at the time of measurement should be clearly and accurately recorded.

F. Remember that nylon or cotton lines or ropes used for any depth related measurements should be recalibrated periodically during the sampling period. Metal clips are recommended for marking depth intervals on lines or ropes, although lines or ropes can also be marked using a pen with permanent ink (e.g., Sharpie).

SOP 5. Water Chemistry, Dissolved Organic Carbon, Dissolved Oxygen Version 1.00

Revision History Log:

Previous Version Number	Revision Date	Author	Changes Made	Reason for Change	New Version Number

Recommended references:

American Public Health Association (APHA), 1998, Standard methods for the examination of water and wastewater (20th ed.): Washington, D.C., American Public Health Association, variously paginated.

Lind, O.T., 1979, Handbook of common methods in limnology (2nd ed.): The C.V. Mosby Co., London.

Wetzel, R.G., and Likens, G.E., 2000, Limnological analyses (3rd ed.): Springer, New York, 429 p.

Water samples are collected for measuring a suite of water chemistry variables. How collected water is processed and maintained in the field depends on the variables of interest for which the water was collected. Since the importance and significance of a suite of variables being measured through time can change relative to changes in sampling objectives or site dynamics, a periodic review of the relevance of the variables being measured should be conducted.

A. **Preparation of Filters, Amber wide-mouth Nalgene® high-density polyethylene (HDPE) and Glass Bottles, and Filtration Apparatus:**

1. Filters

   a. Prepare a relatively contaminant-free space for cleaning and drying the cleaning supplies and sample-collection and sample-processing equipment. This space should also be free of methanol and other organic compounds that potentially could contaminate apparatus to be used for the collection of dissolved organic carbon samples.

   b. **1.2 μm glass-fiber filters (Whatman GF/C)** used to filter water for the analysis of nutrients and total dissolved solids should be pre-washed in 500–1,000 mL of deionized water and dried in a drying oven at 60°C before use in the field.

   c. **0.7 μm glass-fiber filters (Whatman GF/F)** uused to filter water for the analysis of dissolved organic carbon should be combusted in a muffle furnace at 500°C for 4 hours. Filters should be placed individually in aluminum foil envelopes with 3 sides folded closed and the dull side of the aluminum facing inward. The fourth side of the envelope should be open to allow for the escape of gases during ashing. DO NOT combust filters completely enclosed in an aluminum foil envelope or at temperatures above 500°C. Filters should be stacked in the furnace for ashing, and the fourth side of the envelope closed after ashing is completed. (Part of these instructions are from Standard Operating Procedure for the Sampling of Particulate-Phase and Dissolved-Phase Organic Carbon in Great Lakes Waters, Grace Analytical Laboratories, 1994, available at http://www.epa.gov/glnpo/lmmb/methods/pocdoc2.pdf).

   d. Pre-washed and combusted filters should always be handled with clean forceps.

   e. **If a park has previously sampled lakes using a filter with a specific pore-size, and the park would like to compare new water sample data with the old water sample data, then the filter pore-size previously used for water sample collection can be used for the collection of new water samples.**

2. Bottles

a. Amber wide-mouth Nalgene® HDPE bottles, amber glass bottles, their lids, glass Wheaton bottles, and all filtering equipment should be acid-washed prior to use. (***Note:*** *The aluminum liners of the amber glass bottle lids should be removed prior to acid-washing.*) Bottles should be filled with 0.5 N hydrochloric acid and allowed to sit for 24 hours. Lids and filtering equipment should be soaked for 24 hours in 0.5 N hydrochloric acid. The bottles are then emptied and filled with deionized water and allowed to sit for 24 hours. Lids and filtering equipment are soaked in deionized water for 24 hours. The deionized water rinse is repeated. Bottles are then emptied, lids and filtering equipment are removed from their soaking tub, and rinsed continuously for 8 minutes with deionized water. This can best be done in a dishwasher connected to a deionized water source. Bottles, lids, and filtering equipment are allowed to air-dry.

b. After being acid-washed, the amber 60 mL glass bottles for dissolved organic carbon samples and the aluminum liners of the bottle lids are placed into a muffle furnace and heated at 475°C for at least 8 hours. Bottles should be allowed to cool to room temperature. The cooling process should not be hurried as cooling too quickly can cause bottles to crack. Cracked bottles should be considered contaminated and discarded.

c. Lids should be secured to the cleaned amber Nalgene® and amber glass bottles and these bottles should not be opened until used in the field.

B. Collection of the Water Sample:

1. Water samples can be collected using a horizontal Alpha® Water Bottle or a vertical Van Dorn-style water sampler. In shallow systems use of the Alpha® Water Bottle is recommended.

2. Prior to collection of the water sample(s) the collection bottle or sampler should be washed profusely with water from the collection site.

3. All water samples should be collected at the deepest point of each site. The collection bottle or sampler is attached to the end of a line marked at 0.5 m intervals and lowered to a depth that is at the midpoint of the maximum depth for each site. If the site is *thermally stratified*, then two additional samples should be collected: one just below the water surface and one at 1 m above the bottom. See SOP 3 J for the definition of *thermally stratified*.

4. A weighted messenger is attached to the line and upon being released by the technician is delivered to the collection bottle or sampler closing trigger mechanism. This procedure and the action of the messenger triggers the closing of the collection bottle or sampler end plugs into the end openings.

5. The water-filled collection bottle or sampler is then retrieved to the surface, placed into the inflatable boat, and returned to shore for processing.

C. Collection of the Water Sample at Sites < 1.5 m Maximum Depth (Girdner and Larson, 1995):

1. At sites with maximum depth < 1.5 m, water samples should be collected from shore using a wide-mouth 1 L HDPE bottle attached to the end of a telescoping pole.

2. The screw-on-lid of the 1 L HDPE bottle should be removed. The bottle opening is then inverted as the bottle is lowered to approximately mid-depth in the water column near the center of the site.

3. The bottle is then slowly turned so that the opening faces the surface. This action will allow the bottle to fill with water.

4. The bottle is then carefully lifted to the water surface and the water-filled bottle is retrieved to the shore for processing. The lid for the 1 L HDPE bottle should be secured upon retrieval.

5. Water column depth of the sample is visually estimated and recorded on the sample storage bottle and the appropriate data sheet.

D. Field Processing Unfiltered Water Samples:

1. Variables to be measured using unfiltered water: *alkalinity (ANC), conductivity, and pH.*

2. **A minimum of 250 mL of unfiltered water should be collected per site, although 500 mL of sample is preferred.**

3. When the water collection bottle or sampler is brought to and secured onshore, the unfiltered water sample is poured into an acid-washed 250 or 500 mL wide-mouth amber HDPE sample storage bottle.

4. The sample collection bottle should be pre-rinsed twice with 25–30 mL of sample before the final sample is collected into the bottle.

5. Unfiltered water should be poured into the sample storage bottle up to the rim of the bottle prior to securing the bottle lid and be extremely careful not to introduce foreign particulate material into bottle during this process.

E. Field Processing Filtered Water Samples:

1. Variables to be measured using filtered water samples: *nitrate-N, ammonia, total phosphorus, total Kjeldahl nitrogen, sulfate, total dissolved solids (TDS), as well as cations and anions (e.g., Na$^+$, K$^+$, Ca^{2+}, Mg^{2+}, Cl$^-$) if desired.*

2. **A minimum of 500 mL of filtered water should be collected per site.**

3. When the water collection bottle or sampler is brought to and secured onshore a technician can begin the process of filtering the water sample.

4. The collected water is filtered through **a pre-washed 1.2 μm glass-fiber filter (Whatman GF/C) (however, see SOP 5 A 1 e)**, using a 500 mL Nalgene® polysulfone filtering setup (stand-alone or bottletop unit) with hand pump, or a Nalgene® polysulfone syringe-type filter holder.

5. Always place filter in filtering apparatus using clean forceps.

6. Before collecting the sample, pour 25–30 mL of sample into the top chamber of the stand-alone filtering unit or the single water holding chamber of the bottletop unit . Swirl to rinse sides of chamber and discard water. Repeat. If using a 60 mL syringe with syringe-type filter holder, flush 25–30 mL of sample water through syringe. Repeat.

7. Pour another 25–30 mL of sample into top chamber of filtering apparatus and, using hand pump, filter water into lower chamber of filtering apparatus (if using a stand-alone unit); or into an acid-washed wide-mouth amber HDPE sample collection/storage bottle if using a bottletop unit or syringe-type filter holder. Swirl to rinse bottom chamber or sample bottle and discard. Repeat.

8. **NEVER POUR OR RINSE BOTTOM CHAMBER OF STAND-ALONE UNIT OR SAMPLE COLLECTION BOTTLE WITH UNFILTERED WATER. USE ONLY FILTERED WATER TO RINSE THE BOTTOM CHAMBER AND SAMPLE COLLECTION BOTTLE.**

9. If using a stand-alone filtering unit, fill top chamber with 250–500 mL of sample water and pump water through filter. When water has been filtered, pour a small amount (i.e., 25–30 mL) of the filtered water into an acid-washed 250 or 500 mL wide-mouth amber HDPE sample storage bottle. Rinse bottle and discard water. Repeat.

10. Fill sample storage bottle completely to the rim with filtered water prior to securing the bottle lid and be extremely careful not to introduce any particulate material into bottle during this process.

11. Take filter apparatus apart and discard filter.

12. Before and as soon as possible after field use, the filtering unit should be washed in 0.5 N hydrochloric acid, profusely washed and rinsed with low-carbon deionized water, and allowed to air-dry without contamination by dust.

F. Water Sample for Dissolved Organic Carbon (DOC):

1. Water for the DOC sample should be collected from 1 m below the surface at the deepest point of the site.

2. After collecting the water sample, follow the procedure in Section E (3–8) above, except use a **pre-washed 0.7 μm glass-fiber filter (Whatman GF/F)**.

3. Pour small volume (i.e., ~10 mL) of filtered water into a 60 mL amber glass bottle. Rinse bottle and discard water. Repeat twice.

4. **Fill rinsed 60 mL amber glass bottle completely to the rim with filtered water** and place aluminum-lined cap on bottle and secure.

5. Analyses to be completed using this water sample include:

   a. Total DOC.

   b. Attenuation.

   c. Terrestrial fraction. The type of sample processing and analysis should be determined by the objective(s) related to the collection of this sample.

G. Dissolved Oxygen Samples:

1. The following procedure has been reproduced in part from:

American Public Health Association, 1992, Standard methods for the examination of water and wastewater (18th ed.): Washington, D.C., American Public Health Association, section 4500-0 C, p. 4-100.

2. **The chemicals used in this procedure are initially hazardous. Special care should be used when performing this procedure and all technicians should wear latex gloves and eye protection. A copy of the MSDS for each chemical should be available at all times.** The final solution used for the measurement of dissolved oxygen is relatively innocuous. However, this solution should probably be transported back to the laboratory for disposal in a treated wastewater disposal system.

3. *At the monitoring site*, fill a 300 mL Wheaton bottle with water collected from 1 m above the bottom at the deepest point of the site using the collection bottle or sampler. If the site is *thermally stratified* (see SOP 3 J for definition) then a water sample should also be collected from just below the water surface. Attach a tygon tube to the outlet nipple of the collection bottle or sampler and fill the Wheaton bottle with water.

    a. The water in the Wheaton bottle should be completely exchanged at least twice during the process of filling the bottle. This is accomplished by determining the time required to completely fill the Wheaton bottle with water and then allowing water to overflow from the bottle for the same amount of time.

    b. After the bottle has been filled to overflowing, the glass stopper-lid should be placed into the bottle opening and the contents of the bottle should be inspected for air bubbles. **The bottle should contain no air bubbles**. If there are air bubbles in the bottle the filling process should be repeated.

    c. After the bottle has been properly filled, 1 mL of $MnSO_4$ (Manganous sulfate) solution and 1 mL of alkalai-iodide-azide reagent should be added to the bottle.

    d. The pipet tip used to add these chemicals to the bottle should not be dipped into the sample water. The pipet tips should be placed just above the surface of the sample when adding these chemicals.

    e. After adding the chemicals the bottle glass stopper-lid should carefully be put in place without introducing air bubbles to the sample.

    f. The sample should then be mixed by inverting the bottle several times. This procedure will create a flocculent precipitate.

    g. The bottle should then be placed in an upright position and the precipitate should be allowed to settle to approximately half the bottle volume.

    h. Steps f and g should be repeated.

    i. After the precipitate settles, add 1 mL concentrated H_2SO_4 (sulfuric acid) to the sample.

    j. Restopper the bottle and invert repeatedly until the precipitate has been dissolved.

4. At this point the fixed sample (i.e., after steps i and j above) can be titrated at the monitoring site or returned to the laboratory for titration. The titration steps are:

    a. A sample volume of 201 mL should carefully be poured into a beaker and titrated with 0.025 M $Na_2S_2O_3$ (sodium thiosulfate) solution until the sample becomes a pale straw color. The 0.025 M $Na_2S_2O_3$ should be titrated into the sample by placing the $Na_2S_2O_3$ into a 1,000 mL buret with stop-cock attachment. The initial volume of $Na_2S_2O_3$ poured into the buret should be recorded.

    b. A few drops of starch solution are then added to the sample. The addition of the starch solution will turn the sample blue.

    c. Titration of the sample with 0.025 M $Na_2S_2O_3$ should then be continued until the blue color disappears and the sample clears. This is the end point of the titration. The amount in milliliters of 0.025 M $Na_2S_2O_3$ used in this process should be recorded.

    d. If the end point is overrun (i.e., the sample again begins to appear discolored), the sample can be back-titrated with 0.0021 M bi-iodate solution added dropwise until the sample clears.

    e. To calculate the concentration of dissolved oxygen (DO) in the 201 mL sample, 1 mL of 0.025 M $Na_2S_2O_3$ titrated = 1 mg DO/L.

H. Special Considerations:

1. **The water collection bottle or sampler should be preconditioned before sampling additional sites by profusely rinsing with water from the site to be sampled.**

2. The volume of water collected into storage bottles for unfiltered and filtered water samples should be based on the amount of sample required by the laboratory to process and analyze the samples. Amounts between 125–1,000 mL have been requested by various laboratories. These volumes should be decided on prior to the field season and collection of samples.

3. All sample storage bottles should be equipped with labeling tape on which is recorded sample site identifier, sample number, sample date, sample depth, and sample type (i.e., unfiltered water sample, filtered water sample, or DOC). If possible, this should be done before field collection of samples. This information should also be recorded in the monitoring project field book and on the appropriate sample collection datasheet(s).

4. All samples should be inspected to make certain that they are free of floating or suspended debris.

5. Direct handling of the filter, apparatus surfaces that house the filter, and interiors of the water sampler and sample bottles should be avoided. Filters should be handled only with forceps.

6. Prior to sampling and during transport between sites, the filter apparatus and filters should be stored in clean Ziploc® bags to avoid contamination. In addition, filters should be stored in aluminum foil before being placed in Ziploc® bag.

7. Sample storage bottles should be filled completely to eliminate trapped air in the bottles before lids are securely fastened. Trapped air in sample bottles could affect and alter the chemistry of the sample and subsequent analytical results.

8. No preservatives are added to water samples.

9. While in the field, water samples should be kept as cool as possible and shaded from direct sunlight. Bottles containing water samples can be placed into opaque nylon stuff sacks and either buried in snow, if available, or placed in a shaded area of a stream or lake.

10. Unfiltered and filtered water samples should be frozen as soon as possible after leaving the field. Prior to placing water samples into a freezing unit, a small quantity of water should be poured from each sample bottle to permit water expansion during the freezing process and prevent rupturing sample bottles. Generally, sample bottles should be emptied down to the bottle shoulder prior to freezing. Samples should be stored frozen and in the dark and transported frozen to the facility where analyses will be conducted.

11. *DOC samples should not be frozen*. Rather, upon returning from the field, samples should be stored in a refrigerator at 4°C until they can be delivered to the laboratory for processing (preferably no longer than 28 days after collection). Samples should also be kept in the dark at all times.

12. If processing total dissolved solids (TDS) samples in a park laboratory is preferred over sending the samples to a separate laboratory for processing, details concerning the processing of TDS samples can be found in Prepas and Trew (1983), and in APHA (1998).

13. If water chemistry samples are to be processed by a third party, the processing laboratory should be certified for this purpose. Choice of the certified laboratory will be left up to park monitoring project supervisors; however, the same laboratory should be used by all network participants.

SOP 6. Chlorophyll-*a* Concentration and Periphyton
Version 1.00

Revision History Log:

Previous Version Number	Revision Date	Author	Changes Made	Reason for Change	New Version Number

Chlorophyll-*a* concentration can be viewed as a general expression and surrogate measure of algal biomass at monitoring sites. Water samples for measuring chlorophyll-*a* concentration can be collected in conjunction with other water sampling (e.g., collection of water for water chemistry analysis).

A. Collection of the Water for Chlorophyll-*a* Concentration Sample:

1. Water to be filtered for chlorophyll-*a* concentration samples should be collected using the procedure described in SOP 5 B or C.

2. Water for this sample can be collected at the same time water is collected for filtered water chemistry samples, however the filter sizes for water samples for nutrient analysis (1.2 μm) and chlorophyll samples (0.45 μm) are different.

B. Filtering the Water for Chlorophyll-*a* Concentration Sample:

1. A minimum of 500 mL of water should be filtered through a *0.45 μm membrane filter*, using either a 500 mL Nalgene® polysulfone filtering setup (stand-alone or bottletop unit) with hand pump, or a Nalgene® polysulfone syringe-type filter holder. *The filter does not need to be prewashed or ashed.*

2. Place filter in filtering apparatus using forceps.

3. Pour 25-30 mL of sample into top of filter chamber and swirl to rinse sides of chamber and discard water.

4. Using a graduated cylinder, fill top of filter pod with 500 mL of sample water and filter the water. The hand pump pressure gauge should not register higher than 5 PSI.

5. After filtering 500 mL of water, remove filter from filtering apparatus, fold in half, blot off excess water, and place into a 25 mL plastic scintillation vial.

6. Scintillation vial should be wrapped in aluminum foil, placed into an opaque stuff sack, and kept cool until returned to the laboratory.

7. Sample should be labeled with site identifier, sample number, sample collection date, depth at which water was collected, and volume of water filtered. This information also should be recorded on the appropriate data sheet.

8. Upon returning from the field, the scintillation vial containing the sample filter should be placed in a freezer and kept frozen until the sample is processed in the laboratory.

C. Processing Chlorophyll-*a* Concentration Sample in the Laboratory:

Total chlorophyll-*a* concentration of collected samples is determined by the fluorometric procedure described and outlined in APHA (1998).

D. Epilithic Periphyton Sampling:

Periphyton refers to the community of micro-organisms (algae and associated bacteria, fungi, protozoans, rotifers, and other micro-organisms) attached to or on solid submerged surfaces (e.g., rocks and wood) generally above the depth of light extinction (Britton and Greeson, 1987). Epilithic refers to periphyton attached to or on the surfaces of submerged rocks. Several highly recommended references include:

Britton, L.J., and Greeson, P.E. [eds.], 1987, Methods for collection and analysis of aquatic biological and microbiological samples: U.S. Geological Survey Techniques of Water-Resource Investigations, Book 5, Chap. A4, p. 131-144, available online at http://water.usgs.gov/pubs/twri/twri5a4/.

Porter, S.D., Cuffney, T.F., Gurtz, M.E., and Meador, M.R., 1993, Methods for collecting algal samples as part of the National Water-Quality Assessment Program: U.S. Geological Survey Open-File Report 93-409, available online at http://water.usgs.gov/nawqa/protocols/OFR-93-409/alg1.html.

Morin, A., and Cattaneo, A., 1992, Factors affecting sampling variability of freshwater periphyton and the power of periphyton studies: Canadian Journal of Fisheries and Aquatic Sciences, v. 49, p. 1695-1703.

Vinebrooke, R.D., and Leavitt ,P.R., 1999, Phytobenthos and phytoplankton as bioindicators of climate change in mountain ponds and lakes: Journal of the North American Benthological Society, v. 18, p. 14-32.

1. Collection of Periphyton Sample(s):

a. Within the shallow, nearshore area of a monitoring site, five submerged rocks that are at least 0.25 m deep should be located for sampling. (***Note**: The number of rocks collected for sampling should not be less than three and ideally should be ten. However, time and funding constraints will dictate the number of rocks collected and sampled for periphyton*).

b. Rocks should be collected from the major rocky habitats of each site.

c. If a site does not have rocky substrate then periphyton samples are not collected.

d. When a rock is located to be sampled, it should be removed to a processing pan (e.g., 20 × 30 × 10 cm plastic container) located onshore. (***Note:** To minimize human impact in the nearshore area of the lake or pond, rocks to be sampled can be retrieved by a snorkeler.*) The depth of the sample should be recorded.

e. A 125 mL Nalgene® bottle, from which the bottom has been removed, is used to mark on the surface of the rock the area from which the periphyton sample is to be removed.

f. A "stiff-bristled" brush is then used to brush the surface of the rock within the sampling area.

g. After brushing the rock surface for a predetermined amount of time (e.g., 2 minutes), distilled water or filtered water from the site is used to wash the loosened periphyton into the processing pan. This should be done with the technician washing the rock surface while holding the sample rock off of the bottom and above the processing pan. The brush and brush bristles also should be washed and the effluent allowed to fall into the processing pan.

h. Water and suspended periphyton is then poured from the processing pan into a sample collection container (e.g., 250 mL Nalgene® bottle).

i. Processing pan is then rinsed with distilled or filtered water and this effluent is poured into the sample-collection container.

2. **Filtering and Processing the Periphyton Sample(s):**

    a. Because the periphyton sample(s) will be analyzed for chlorophyl-*a* concentration, the collected sample will be filtered similar to water filtered for chlorophyll-*a* concentration analysis.

    b. Filtering of the periphyton sample can be completed in the field or returned to laboratory for filtering. If the periphyton sample(s) is to be processed in the laboratory, the sample(s) should be kept in the dark and as cold as possible until the sample(s) can be filtered.

    c. The periphyton sample collected into the sample container should be filtered through a *0.45-μm membrane filter*, using either a 500 mL Nalgene® polysulfone filtering setup (stand-alone or bottle-top unit) with hand pump, or a Nalgene® polysulfone syringe-type filter holder. *The filter does not need to be prewashed or ashed.*

    d. Place filter in filtering apparatus using forceps.

    e. Fill top of filter pod with the periphyton sample (algae and water).

    f. Filter sample making sure that the hand pump pressure gauge does not exceed 5 PSI.

    g. After filtering the sample, remove the filter from filtering apparatus and place into a 25 mL plastic scintillation vial.

    h. The scintillation vial should be wrapped in aluminum foil, placed into an opaque stuff sack, and kept cool until returned to the laboratory.

    i. The sample should be labeled with site identifier, sample number, sample collection date, depth at which rock was collected, and area sampled. This information also should be recorded on the appropriate data sheet.

    j. Upon returning from the field, the scintillation vial containing the sample filter should be placed in a freezer and kept frozen until the sample is processed in the laboratory.

    k. Periphyton samples will be processed using the fluorometric procedure described and outlined in APHA (1998).

3. **Frequency of Sampling and Location of Processing:**

    a. The frequency of sampling is cost dependent so the number of monitoring sites and samples per site can be adjusted based on available funding.

    b. If funding for sample processing is limited, periphyton sampling could be limited, for instance, to three samples per each Intensive Sampling monitoring site (see Sample Design, table 1, Level 1).

    c. Because the filtering of periphyton samples can be quite time consuming, it is left to the discretion of the monitoring project Principal Investigator as to whether these samples will be processed in the field or returned to laboratory for processing.

SOP 7. Zooplankton (Rotifers and Crustaceans)
Version 1.00

Revision History Log:

Previous Version Number	Revision Date	Author	Changes Made	Reason for Change	New Version Number

A. Collection of Samples at Sites ≥2 m Maximum Depth:

1. Zooplankton samples are collected using a 64-μm mesh Wisconsin-type zooplankton net, with a 20 cm diameter opening and 1 m long from the bottom of the collecting cup to the top of the opening. Net efficiency is assumed to be 100 percent (however, efficiency is more like about 70 percent).

2. The collecting cup is attached to the net and the net is attached to the end of a nylon or cotton rope marked at 0.5-m intervals.

3. The net and collecting cup (i.e., collection unit) are lowered from an inflatable boat (or float tube) to where the bottom of the collecting cup is within 1 m of the bottom of the monitoring site.

4. The collecting unit is then retrieved through the water column to the surface by pulling the rope or line **vertically** at a rate of approximately 0.5 m/s.

5. When a **vertical tow** is completed, each sample is inspected for organic and inorganic debris (e.g., fir needles, pollen, parts of leaves, silt and bottom sediment). Samples with excessive amounts of debris are discarded and the preceding procedures are repeated until an acceptable sample has been collected. Three acceptable **vertical tows** are performed at each site.

6. When a debris-free sample is retrieved to the surface, contents within the net are rinsed into the zooplankton cup.

7. To rinse zooplankton from the net into the cup, the net is slowly lowered into the lake or pond to within inches of the net opening and abruptly retrieved upward out of the water. Care should be taken not to lower the net opening beneath the lake or pond surface.

8. After several rinses of the net, the cup is detached from the net and its contents are again inspected for debris. The amount of water in the cup is then reduced by swirling the contents of the cup so that the water filters out of the cup through the 0.64-μm mesh screens on the side of the cup.

9. The sample of concentrated zooplankton is then rinsed from the cup into a 60 mL HDPE sample bottle using a squirt bottle filled with deionized water or pond/lake water filtered using a drinking-water filter. The water should be carbonated using sodium bicarbonate. This can be achieved by adding a small piece of Alka-Seltzer® to the water in the squirt bottle.

   a. The sample bottle should end up being approximately one-quarter to one-third full with zooplankton and water.

   b. This procedure should be performed for each vertical tow.

10. After the zooplankton in each vertical tow have been rinsed into the sample collection bottle, 95 percent ethanol is added to the sample bottle until the bottle is full.

11. Site identifier, sample number, sample date, and tow length (i.e., distance in meters of water column through which the collecting unit was retrieved) should be marked on the sample bottle label and appropriate data sheet.

B. Collection of Samples at Sites <2 m Maximum Depth:

1. Zooplankton samples are collected from shallow monitoring sites using the same net configuration used to collect samples at deeper sites (SOP 7 A). However, an additional rope or line is attached to the collecting cup end of the sample collection unit.

2. The sample collection unit is pulled **horizontally** rather than vertically through the water of shallow sites.

3. Three replicate **horizontal tows** are performed from shore for each site.

4. Two crew members are required for the collection of these **horizontal tow** zooplankton samples.

5. Crew members are positioned opposite of one another along the longitudinal axis of the site. Together they lower the sample collection unit below the water surface and the crew member holding the rope or line attached to the net opening retrieves the net by pulling the rope or line, marked at 0.5-m intervals, to the shore at a rate of approximately 0.5 m/s.

6. The crew member opposite the crew member retrieving the net makes certain that the sample collection unit remains **horizontal** in the water column and that the collecting cup does not drag on the bottom substrate of the monitoring site.

7. Procedures outlined in SOP 7 A 5–10 should be followed to process and preserve samples collected from shallow sites.

8. Site identifier, sample date, and tow length (in meters) should be marked on the sample bottle label and appropriate data sheet.

9. Water can be collected from ponds that are very shallow using the method described in SOP 5. C (Girdner and Larson, 1995). When using this method, 2 L of water should be poured through the zooplankton net with the collecting cup attached. The sample is then processed the same as samples collected using a flotation device (i.e., SOP 7 A 5-10).

C. Special Considerations:

1. After a site has been sampled, the net, collecting cup, and retrieval rope or line should be thoroughly rinsed (e.g., 2 to 3 times) with water from the site and allowed to dry if possible.

2. Prior to use at another site, the net, collecting cup, and retrieval rope or line should be preconditioned by rinsing 2 to 3 times with water from the new site.

3. Once out of the field, the ethanol in sample bottles should be replaced with a 70 percent ethanol solution.

4. Bottle lids should be checked to make certain that they are securely fastened. In addition, samples to be stored for a lengthy period of time prior to processing should have their lids taped.

SOP 8. Aquatic Macroinvertebrates
Version 1.00

Revision History Log:

Previous Version Number	Revision Date	Author	Changes Made	Reason for Change	New Version Number

I. Option 1:

Sampling for aquatic macroinvertebrates is conducted in major nearshore habitats and from deeper offshore locations in monitored ponds and lakes. Sampling for macroinvertebrates using Option 1 will provide the following basic information: (a) whole site community composition, and (b) taxa presence and relative abundance per site and per major habitat types.

A. **Identification of Major Nearshore Habitats:**

1. Habitats are identified by crew members by either walking or boating the entire perimeter of a site and identifying the dominant substrate types present in an area from the shore to a point offshore where the water depth is approximately 50 cm. Bahls (1991) used a similar classification system for identifying nearshore habitats of mountain lakes in Idaho by recording habitat information at 20-m intervals along the perimeter of each study lake.

2. Substrate components are identified using the Wentworth scale as modified by Cummins (1962). Major habitats are defined by what is estimated to be the most abundant substrate component in an area: silt, sand, gravel, cobble, boulder (small or large), bedrock, organic detritus, fine wood, coarse wood, aquatic vegetation.

3. Combinations of substrate types (e.g., silt-sand, gravel-cobble, silt-coarse wood, etc.) also can be used to identify a habitat type when these substrates are estimated to be co-dominant in an area.

4. Linear shoreline distance is measured for each major habitat of a lake or pond and recorded on a representational shoreline map of the site as well as on the appropriate data sheet. Major habitats should be identified sequentially such as habitat 1, habitat 2, habitat 3, etc.

B. **Nearshore Sampling:**

1. Each major nearshore habitat of a site should be sampled.

2. Quantitative sampling:

   a. The process for sampling within a habitat begins by placing a 1 m^2 PVC-frame on the surface of the habitat-substrate to be sampled. Placement of the frame does not necessarily need to be random.

   b. A crew member, using a standard D-shaped sampling net with a handle ≥ 2 m long captures macroinvertebrates by sweeping the net back and forth through the PVC-framed sample area while disturbing the substrate within the 1 m^2 PVC-frame with his/her foot. The bottom of the net-frame should be held just above the habitat-substrate surface to capture organisms and some substrate in the net during the sweeps. This process should be conducted for from 30 to 60 seconds.

3. Qualitative sampling:

   a. This process is similar to the quantitative method outlined above except that the 1 m² PVC-frame is not placed in the habitat to mark the area to be sampled.

   b. Individual organisms of interest also can be sampled from habitats by either capturing them with the net or by using forceps to remove them from substrates.

   c. The primary purpose of this sampling is to establish a record of the presence of macroinvertebrates at each monitoring site and relate that presence to habitat as defined by dominant substrate(s).

4. Sample Processing: Once the sample has been collected the crew member then returns to the shore to process the net contents in one of two ways:

   a. **Method 1:**

      (A) All contents (i.e., substrate and captured organisms) of the D-shaped sampling net are placed directly into a 1 gallon Ziploc® bag or 42 ounce WHIRL-PAK® sampling bag; or

      (B) Net contents are subsampled prior to placement into plastic bag and preservation;

      (C) If net contents are subsampled, the amount of the subsample needs to be estimated as a percentage of the total sample collected and recorded on the appropriate data sheet;

      (D) The plastic bag is then filled with enough 95 percent ethanol to cover the substrate;

      (E) Sample bags are labeled with site identifier, sample date, major nearshore habitat number, and estimated amount of subsample (if appropriate);

      (F) This information also is recorded on the appropriate data sheet;

      (G) In the laboratory, the contents of the bag are processed by picking through the substrate for organisms using forceps and a dissecting microscope;

      (H) Organisms removed from the substrate are placed into 2 dram vials partially filled with 70 percent ethanol;

      (I) A Rite-in-Rain® label with site identifier, sample date, major nearshore habitat type, and estimated amount of subsample (if appropriate) is placed into the vial; and

      (J) This information also is recorded on the appropriate data sheet.

   b. **Method 2:**

      (A) All contents (i.e., substrate and captured organisms) of the D-shaped sampling net are emptied directly into a rectangular plastic container and macroinvertebrates are picked from the substrate with forceps (*Note: Large debris such as stones and wood should be inspected for organisms and discarded*); or

      (B) A subsample of the net contents is placed into the rectangular plastic container prior to the picking of macroinvertebrates from the substrate;

      (C) If the contents are subsampled, the amount of the subsample needs to be estimated as a percentage of the total sample collected and recorded on the appropriate data sheet;

      (D) Macroinvertebrates picked from the substrate are placed into 2 dram vials partially filled with 70 percent ethanol.;

      (E) A Rite-in-Rain® label with site identifier, sample date, major nearshore habitat type, and estimated amount of subsample (if appropriate) is placed into the vial; and

      (F) This information also is recorded on the appropriate data sheet.

C. Offshore Sampling (optional):

Offshore samples most often are collected from sites that are relatively easy to access. Backcountry or cross-country sites in rugged or difficult to access locations are not suitable sites for offshore sampling using the relatively heavy Ekman dredge.

1. A 6-in. Ekman Dredge is used to collect offshore soft-substrate samples (see Merritt and others, 1984).

2. The open dredge is attached to a rope or line marked at 0.5-m intervals and lowered to the bottom of the site at a deep water sampling location. The dredge is allowed to descend at a rate that will assure that the open end of the dredge will be depressed into the soft-substrate of the site.

3. The crew member then attaches a messenger to the rope or line and allows the messenger to drop to the dredge, triggering the closing of the dredge opening.

4. The dredge is then slowly retrieved to the surface and lifted into a large plastic garbage bag or plastic bucket.

5. The crew member then returns the sample to the shore where it is further processed.

6. The substrate sample can then be processed as described in SOP 8 B 4a or b; or alternatively the sample contents, because they are usually soft-substrate, can be rinsed in a 500-μm sieve. Rinsing of the substrate in the sieve involves dipping the sieve and its contents into the pond or lake and shaking or twisting the sieve in the water. The top open part of the sieve should not be submerged below the water surface. The organisms captured in the sieve during this rinsing process can then be transferred using forceps to a 2 dram vial partially filled with 70 percent ethanol. This process should be repeated until all or a subsample of substrate has been processed. (However, the substrate collected in the sieve during the rinsing process, or a subsample of this substrate, can be collected into a gallon Ziploc® bag, preserved with 95 percent ethanol, and returned to the laboratory for future processing.) When the sample or subsample processing has been completed, a Rite-in-Rain® label with site identifier, sample date, major offshore habitat type, and estimated amount of subsample (if appropriate) is placed into the vial and this information also is recorded on the appropriate data sheet.

7. **Safety Notice: Deployment of an Ekman dredge from an inflatable boat or other flotation device can be hazardous and should be performed with utmost care and caution to avoid puncturing the boat or device while on the pond or lake.**

D. Sample Site Locations:

The location of the sampling site(s) within each major nearshore habitat and for offshore samples should be identified on a representational map of each site. Sample site locations also can be identified by UTM coordinates. This information should be recorded on the appropriate data sheet. During any subsequent sampling visits, the sample site locations can be relocated and resampled if desired.

E. Cleaning Sampling Equipment:

1. In the field after the completion of macroinvertebrate sampling, all sampling equipment should be thoroughly washed with water from the sampling site and allowed to air-dry.

2. Prior to sampling at a new site, all sampling equipment should be preconditioned with water from the new site. Water used for this purpose should be visually inspected for the presence of macroinvertebrates to insure, as best as possible, that sampling equipment is free of organisms prior to the beginning of sampling and sampling between habitats.

3. In the laboratory, all sampling equipment should be washed with detergent, soaked and thoroughly rinsed in water, and allowed to air-dry before the next sampling trip.

II. Option 2: BMI Method (prepared by Reed Glesne, NOCA)

A. Objectives

1. Primary objectives of Option 2 are as follows;

    a. Obtain representative littoral zone samples of pond and lake benthic macroinvertebrate (BMI) communities that:

    (1) Incorporate the heterogeneity of habitat types found in those waters;

    (2) Maximize species richness information;

    (3) Provide semi-quantitative data for development of community metrics.

    b. Define pond and lake reference site groups based on BMI community data, and facilitate comparisons of future test sites with known reference conditions.

    c. Apply results for use in broad scale evaluations of status and trends.

    d. Assure sampling and laboratory analysis efficiency (effort and cost).

    e. Minimize logistical constraints associated with transportation of equipment and samples to and from backcountry sites.

B. Environmental attribute data required for BMI analyses

Environmental attribute information will be collected *a priori* using maps, aerial photos, Geographic Informational Systems (GIS), and during sample collection. This information will consist of both inherent predictor variables and anthropogenic disturbance parameters. Predictor variables are used in the development of 'Multivariate' models to predict the occurrence of taxa in minimally impacted reference sites. Predictor variables represent features generally unaffected by anthropogenic influences because they are used to partition the natural variability among reference sites. Anthropogenic disturbance parameters will be used to place ponds and lakes in disturbance categories and to establish reference site criteria. Additional physical and chemical attributes (explanatory attributes) used in explaining or interpreting variation in biological community data and levels of impairment will be collected.

1. Predictor variables

    a. Variables to be considered in the multivariate predictive model will include catchment area, pond/lake size and depth, elevation, latitude and longitude, shoreline habitat characteristics, outlet presence and flow, glacial influence, water temperature, and conductivity.

    b. GIS applications will be used to determine catchment area, the percent of the catchment occupied by glaciers, pond/lake size and elevation based on 1:24,000 topographic maps in Arc Info 8.0.

    c. Shoreline characteristics will be gathered in the field with the aid of aerial photographs to determine the percent area of the pond/lake perimeter (within the first 20-m contour interval from the shoreline) covered by snow/ice, talus/cliff, meadow, meadow/shrub, and forest.

    d. Mean and maximum depth will be determined following SOP 2.

    e. The number of inlets and presence of outlets will be recorded (SOP 1). Outlet discharge will be measured according to USFS Region 6 protocols (USFS, 2003) using a Swoffer Model 2100 current velocity meter.

    f. Water temperature and conductivity will be measured according to SOP 3 and SOP 5.

    g. Other variables to be considered (found in SOP 1) include Site Basin Aspect and Site Basin Origin.

2. Disturbance parameters

    a. Each lake or pond will be rated for several anthropogenic disturbance factors that occur within the catchments and within the immediate area of each site.

    b. Disturbance factors will include recreational use and infrastructure, landscape condition class, park developments, hydraulic modifications, road density, and abundance of non-native fish (SOP 11).

    c. Other park-specific past land use activities such as mining, grazing, logging etc. will be considered.

    d. Additional disturbance components, disturbance scoring, and development of levels of impairment will be determined by applying the disturbance related field data, and by interpretation of the relationships of physical, chemical and biological data collected, to the various components of the disturbance gradient.

3. Explanatory attributes

    a. Additional physical and chemical attributes used in explaining or interpreting variation in biological community data and levels of impairment will be collected.

    b. During each site visit, a suite of water chemistry and physical habitat attributes will be collected from each pond/lake.

    c. Predominant features of pond/lake habitat will be characterized at each BMI sample site according to substrate type and size (table 3).

Table 3. Lake habitat parameters.

Type	Size
Inorganic	Silt
	Sand
	Gravel (2 to 64 mm diameter)
	Cobble (64 to 256 mm diameter)
	Boulder (>256 mm diameter)
Organic	Fine Particulate Organic Matter
	Detritus (Coarse Particulate Organic Matter)
	Coarse Woody Debris
Vegetation	Emergent
	Floating
	Submerged

d. The percentage of the sample area (see table 4 for sample area) covered by each substrate type and size will be determined by visual estimation and description will be based on 10 percent categories up to 100 percent (e.g., for sample areas with multiple substrate types and sizes the estimate could be something like 10 percent Silt, 10 percent Sand, 30 percent Detritus, 50 percent Coarse Woody Debris = 100 percent; for sample areas with one substrate type and size, for instance Gravel, the estimate would be 100 percent).

e. Water chemistry data including dissolved oxygen profiles, pH, specific conductance, alkalinity, and transparency will be collected at each pond and lake (SOP 4 and SOP 5).

f. Samples will be collected for dissolved organic carbon (DOC), nutrients, selected cations and anions, and chlorophyll-*a* (SOP 5 and SOP 6).

C. BMI sampling methods

1. Sample site location and collection

    a. BMI will be sampled using a semi-qualitative kick-and-sweep technique. The objective of this technique is to sample all habitat types within randomly selected sites within the 1-m depth contour of each pond/lake to be sampled.

    b. Samples are collected by sweeping through suspended bottom material kicked up from the pond/lake bottom or brushed from large woody debris, rocks and vegetation using a 500-μm mesh D-frame kick net.

    c. Sampling effort is constrained by linear distance and time. In order to minimize disturbance, the time and linear distance of sampling is variable depending on pond/lake size.

    d. Number of pond/lake subsamples, linear distance sampled for each subsample, and time constraint for each subsample by pond and lake surface area strata are shown in table 4.

Table 4. Pond and lake sample effort by lake size strata.

Lake surface area (ha)	No. of subsamples	Linear dist. (m) per subsample	Time (min) per subsample
< 0.2	5	3	1
0.2 - < 0.8	5	5	2
0.8 - < 4.0	5	10	3
> 4.0	5	20	4

e. In table 4, the total linear distance sampled is approximately 5 to 15 percent of the total shoreline distance based on a shoreline development ratio of 1.0.

f. At each pond/lake, the five subsamples are taken from randomly distributed locations within the 1-m depth contour sampling area.

g. Pond/lake shorelines are divided into approximately 20–25 numbered segments on an aerial photograph and five random numbers are selected to represent the segments that would be sampled. (*Note*: *An oversample of two or three additional random numbers are drawn at ponds/lakes that may have areas that cannot be sampled.*)

h. Within each sample segment, the center of the segment is located and the sample area is designated as the area within the 1-m depth contour, and one-half of the distance in table 4 is sampled on either side of the segment center point.

i. Each area is sampled with the kick net for the total minutes listed in table 4.

2. Field sample processing

    a. Once collected, each subsample will be elutriated in a white photographic tray or bucket and washed through a hand-held 500-μm mesh net.

    b. Any organic matter to be discarded will be inspected for specimens.

    c. This process serves to separate the coarse organic matter from the sample insuring better specimen preservation.

    d. Subsamples will be individually double bagged in WHIRL-PAK® bags, labeled (duplicate labels—one in each bag) and preserved with 75 percent ethanol.

    e. Labels will include site name and subsample number (and subsample partition number if necessary), sample date and initials of collector.

    f. All subsample location numbers will be recorded on a field map along with other pertinent information from other SOPs.

D. Laboratory procedures

1. Sample storage

    a. All samples from each site will be immediately refrigerated (4°C) as they come in from the field.

    b. A sample log will be continually updated concerning dates and disposition of samples (i.e., refrigerated, sorted, specimen identification, data entry, and location of processed voucher samples).

2. The following subsampling procedures were adapted from Cuffney and others (1993) and the Utah State University Bug Laboratory [*Note: The full process described in Cuffney and others (1993) will be used for a subset of lakes as part of QA/QC procedures for looking at sampling variation among the five samples taken at each pond/lake. For the purpose of saving time and cost we will consider compositing all 5 samples for most ponds/lakes, and then processing the composited samples using the Cuffney and others (1993) procedures to get a minimum of 500 organisms (compared to the 300 organisms per subsample described below) for the entire composite pond/lake sample.*]

    a. The goal of the subsample processing procedures is to remove an unbiased, random representation of macroinvertebrates from each pond/lake sample.

    b. For the purposes of this SOP option, a sample is identified as the location where the kick-and-sweep was done.

    c. Each pond/lake should have five samples. The analysis of these samples will take into account the composition of macroinvertebrate species found at each site and their distribution throughout the pond/lake sampled. Therefore, it is important that throughout processing each sample be kept in a separate labeled vial and (or) whirl-pak.

    d. All labels, at a minimum, should contain a unique identifier for the sample (Pond or Lake Name/Number and Sample Number), the date, who collected the sample, and the acronym for the Park where the collection is from.

    e. Labels should be written on Rite-in-Rain® paper with pencil.

    f. If more than one person is working in the laboratory, all containers (whirl-paks, cups, and vials) must be labeled to avoid confusion.

    g. The size of the subsample is a minimum of 300 individual organisms from each sample (for all 5 of the sample site locations at a pond/lake).

    h. These organisms must be taken from a minimum of one-quarter of the sample (it is possible to pick more than 300 organisms in order to meet this criterion).

3. Subsample processing

    a. Individually rinse each sample into a 500 µm mesh cone and pour remaining contents into a single container with enough water to suspend the sample material.

    b. Thoroughly mix the material together.

    c. Place a sheet of 500 µm mesh into a tray containing approximately 3 cm of water to facilitate the even distribution of sample material. The mesh should be divided in half and numbered with a permanent marker to facilitate accurate identification of subsample size.

    d. Evenly distribute the material from the cone on the surface of the mesh and gently lift out of the water.

    e. Use a knife or spatula to separate the sample along the marked line.

    f. Flip a coin to determine which half of the sample is to be processed; heads = #1, tails = #2.

    g. Keep the portion of the sample to be processed on the mesh. Place the other half into a container filled with water.

    h. Label this portion of the sample 50 percent, signifying that it contains one-half of the original combined sample. If you judge that you will start picking with less than 50 percent of the sample, place the mesh back in the water, and re-float the material to evenly distribute it.

    i. Re-flip the coin and divide the portion in half again.

    j. Place the material you are not going to immediately sort into a different container, and label with the split percentage (e.g., 25 percent).

    k. Repeat this process until it appears that approximately 300 organisms remain on one-half of the mesh. The goal is to remove at least 300 organisms.

l.   Place the split to be sorted into a container partially filled with water and decant a portion of this material into a shallow bottomed sorting tray. Once you start a split, you must finish it in its entirety.

m.   Remove all organisms from the sorting tray using a magnifying (3-10×) lens and place into vials with a preservative (75 percent ethanol). If you are introducing small droplets of water into the vial with specimens, you will need to decant some of the fluid in the vial and refill with the concentrated preservative.

n.   Organisms should be sorted into the following categories: (A) "worms" (flat worms, oligochaetes, nematodes, and Nematomorpha); (B) midges; (C) all other insects; and (D) Mollusks. Not all these categories will be represented in every sample. **Oligochaetes should be picked and preserved but are not counted as part of the 300 organisms total. These worms tend to break into several pieces during the preservation process.**

o.   If less than 300 organisms were removed in the previous steps, then repeat the process outlined in the previous steps until the minimum 300 organism target is obtained (minus oligochaetes).

p.   Before discarding any debris it must be checked by another person to ensure that no organisms are being lost. If you are processing samples by yourself put the debris in a whirl-pak with preservative to be held until a second party can look through the debris.

q.   When you have removed 300 organisms, spread the remaining unpicked sample evenly throughout the sorting pan.

r.   Systematically search the pan and remove any organisms that you think you might not have found in your split samples. Look for new or different organisms having different colored head capsules, body postures, sizes or coloration. **Remember: when in doubt pick it out.** It is better to pick up duplicates than to miss something. Keep these specimens in a separate vial that is labeled as qualitative.

s.   After completing the quantitative and qualitative picking, the remaining debris can be discarded.

t.   On the laboratory data sheet record the following information:

(1)   unique identifier for the sample

(2)   acronym for the Park where the sample was collected

(3)   pond/lake name/number

(4)   sample number

(5)   proportion of the original sample that was subsampled

(6)   date sample was collected

(7)   date sample was processed

(8)   initials of the people who collected the sample

(9)   initials of the person who picked the sample

(10)   number of vials the specimens were put into

4.   Specimen identification

a.   All organisms will be identified to lowest practical taxonomic level, usually genus, counted, and transferred to 75 percent ethanol.

b.   All generic level identifications follow Merritt and Cummins (1996). Various references, as outlined by Plotnikoff and White (1996), also are used for family, genus, and some species level identification.

c.   Specimens that are in good condition or not already represented in the Park macroinvertebrate reference collection will be archived in the collection. Also, a regional macroinvertebrate reference collection should be developed.

d.   Secondary identifications on 10 percent of the samples collected will be performed by taxonomists that have considerable experience with taxa found in the Pacific Northwest.

e.   Problematic taxa will be sent to various experts for verification prior to inclusion in the dataset and reference collections.

E. Data analyses

'Multimetric' and 'Multivariate' methods will be used for evaluation of BMI data. These methods have been proven to be effective in detecting impairment attributed to various perturbations and rely on collection of biological community data from a range of reference sites.

The 'multimetric' approach uses a variety of BMI metrics (representing functional, compositional, life history and pollution tolerance characteristics of the communities) that are determined for each site and measured for their performance in detecting change along the disturbance gradient. Metrics that are capable of discriminating between different levels of disturbance are included in the final 'multimetric' Index of Biological Integrity (IBI). Additionally, numerous other metrics will be individually evaluated at sites where trend data are being collected.

Data analysis protocols for the 'Multimetric' approach are described by Plafkin and others (1989), Hayslip (1993), Barbour and others (1996), Barbour and others (1997), Karr and Chu (1997), and Stribling and others (2000). A number of potential benthic metrics have been described by Barbour and others (1997), U.S. Environmental Protection Agency (1998), and Lewis and others (2001).

The 'Multivariate' or 'Predictive Model' approach will use BMI data from a set of unimpaired sites, at each park unit, that represent a wide range of environmental variation (stratified initially by watershed, elevation, and pond/lake size class). Pond and lake sites are classified into groups based on similarity in their species composition, using ordination or clustering methods. A method is then required to match a test site to the appropriate reference group. A discriminant function based on environmental attribute parameters (independent of change related to human disturbance) of the reference site data is used to predict group membership of test sites. If a test site can be associated with a group of reference sites, representing unimpaired conditions, then the reference site data can be used to predict the fauna expected at the test site. Deviation in the expected vs. observed frequencies of occurrence of taxa between the reference data set and the test site data set are used to evaluate impairment. The sensitivity of this method can be determined by comparing the reference data sets with matching test sites of known impairment. 'Multivariate' data analysis methods are given in Moss and others (1987), and Barbour and others (1997).

SOP 9. Aquatic Amphibian Sampling and Surveys
Version 1.00

Revision History Log:

Previous Version Number	Revision Date	Author	Changes Made	Reason for Change	New Version Number

This SOP gives step-by-step instructions for performing visual encounter surveys for pond-breeding amphibians and details the methods for quantifying habitat variables at the survey site. The goal is for each pond/lake to be visited at least two times during the snow/ice free field season (usually late-June through mid-September) when aquatic life stages of native amphibians are likely to be present and detectable. Sampling some ponds/lakes only once will increase the error associated with the estimate of the proportion of sites occupied (PAO, see SOP 10). Sampling some ponds/lakes only once also requires the assumption that detectability patterns at these sites are the same as at the other sites.

A. Materials:
These materials are needed to conduct the survey:

1. Map of study area that indicates the locations of survey ponds/lakes and their site ID numbers.

2. Data notebook with blank datasheets copied on waterproof paper or a Personal Digital Assistant (PDA) with a water resistant case and memory chip for backing up data on the PDA in the field.

3. Pencils and permanent markers.

4. Thermometer for measuring air and water temperatures.

5. Dip nets (long-handled, 3-mm mesh).

6. Re-sealable plastic bags.

7. Waders (optional if you do not mind getting wet) and wading shoes.

8. Metric ruler (to measure amphibian lengths).

9. Hand lens (for examining features on larval salamanders and tadpoles).

10. Compass (set to correct declination).

B. Procedures:

1. **Prior to Starting Tour:**

    a. Determine which sites will be visited during the tour, and make maps to take into the field that indicate the locations and site ID numbers of the sites to be surveyed.

    b. Make sure that you have all the materials necessary to conduct the surveys planned for the tour. Unmapped ponds/lakes that are encountered will also be sampled, so be sure to have enough materials to survey all sites.

2. **Upon Arrival at a Site:**

    a. Verify that you are at the correct site by checking the map and your GPS unit.

    b. A new datasheet should be used for each discrete body of water. If ponds/lakes are connected with channels or are part of a wet meadow, the ponds/lakes are not discrete and should be treated as one site. Data from non-discrete sites should be recorded on one datasheet and treated as one site. A complex of very small ponds separated by no more than 10 m can be treated as non-discrete.

3. **Filling out the Datasheet or PDA Data Forms:**
   A sample datasheet and definitions for data fields
   are provided in Appendix VII. Instructions for
   completing data forms on PDAs are provided in
   SOP 11.

   a. Begin filling out the datasheet for the site to
      be surveyed. See the field descriptions for a
      detailed description of each data field. Be sure to
      write "NA" or strike through boxes that are not
      applicable to a survey instead of leaving them
      blank. A brief note on why missing data were not
      collected should be included in the notes section
      (e.g., "animal escaped before measurements
      taken").

   b. Record the time that you start working at the site
      in the start time box.

   c. Record as much of the basic data about the site
      as you can. This should include the site ID,
      drainage, location and weather data for the site.

4. **Survey Techniques:**
   The rest of the steps in the sampling protocol are not
   order-dependent and may be completed in any order.

   a. **Survey Site Map:** On page 2 of the datasheet,
      draw a map of the pond/lake perimeter in the
      box. Use a compass to make sure your drawing
      is oriented to the North arrow on the map.
      Indicate habitat details such as large areas of
      emergent vegetation or log jams. Use symbols
      to indicate where survey activities are conducted
      (e.g., temperature measurement, GPS location,
      photograph location).

   b. **GPS coordinates:** Follow the manufacturer's
      directions for operating the GPS unit that you
      are using. Record decimal degree coordinates
      in the WGS84 datum. Find a location along the
      shoreline where the GPS unit receives a strong
      signal and indicate that location on the survey
      site map (see above). Continue to log points
      at that location for 1 to 2 minutes. Record the
      decimal degrees and an estimate of error (if
      your GPS unit can supply an error estimate).

Proportional dilution of precision (PDOP) is
the preferred measurement of GPS accuracy. A
lower PDOP indicates a smaller error. Try to take
GPS measurements when the PDOP is low (i.e.,
below 6). Satellite geometry and atmospheric
conditions affect PDOP values and charts can
be prepared that indicate times of lower PDOP
values. It may be advantageous to plan sampling
during times of lower PDOP values, but this is
not always possible. If the coordinate file was
saved, indicate the file name on the datasheet.
Do not use the elevation returned by the GPS
unit because these readings tend to be inaccurate.
Instead use a topographical map to determine the
elevation.

   c. **Visual Encounter Survey:** Surveys generally
      will be conducted by crews of two before or
      during the aquatic habitat characterization.
      Surveys will usually be conducted during the
      day. However, at sites where no/few animals
      are detected during day surveys, night surveys
      also can be conducted. At sites where multiple
      surveys are to be conducted, either during the
      day and/or at night, the additional survey(s) can
      be conducted immediately after the first survey,
      i.e., no elapsed time between surveys.

   (1) One surveyor will walk along the shoreline,
       while another will wade. At sites where
       the potential of negative nearshore habitat
       disturbance is a concern, the wading-
       surveyor can either snorkel or use a
       floatation device.

   (2) Surveyors walk/snorkel/float the perimeter
       of the wetland continuously searching for
       amphibians and stopping every 10 m, or
       for no more than 20 evenly spaced stations,
       to search more carefully for embryos and
       larvae. Vegetated areas and other habitats
       likely to harbor amphibian embryos and
       larvae that fall between stations should be
       intermittently searched with dip nets.

(3) A rough procedure is used to determine the locations of stations around the pond/lake perimeter where the more focused searches for embryos and larvae are conducted.

    (a) First, surveyors decide how many stations up to 20 to use for the survey (*Note: The maximum of 20 stations is typically used at sites greater than 1 ha*).

    (b) Second, surveyors estimate the distance needed between stations to obtain the desired number of evenly spaced stations located along the entire pond/lake perimeter. This estimate will be very rough and the distance between each station does not have to be equal; rather just relatively equivalent.

    (c) As the surveyors progress around the site, they should visually estimate where the next station will be and pick a landmark of some sort to identify the next station location. This procedure is repeated from station 1 to station 2 to station N.

    (d) At the estimated ¼, ½ , and ¾ points along the site perimeter, the surveyors should confirm that they have done ¼, ½ , and ¾ of the stations that they plan to complete at the site. If they are above or below schedule for the number of stations, they should have completed by the estimated ¼, ½, and ¾ points along the perimeter, the surveyors should adjust the distance between the remaining stations accordingly.

    (e) If good/reliable information is available about the perimeter length of the site (e.g., based on a map or GIS information), the actual distance between stations could be calculated; and at the site the distance between stations could be paced off by the onshore surveyor.

    (f) The main point of this discussion of station location identification is to encourage surveyors to develop a system for identifying station locations that prevents the surveyors from simply walking along the perimeter and suddenly deciding "how about here?" for a station location.

(4) At each station, surveyors search for amphibians in hidden microhabitats and, if visibility is limited, make sweeps with a dip net to capture individuals to verify the species identification. If there are movable cover items at a station, move them to see if there are any amphibians beneath the items. The search area should be about 2 m² around each station, extending from the shoreline toward the center of the pond/lake.

(5) **Aquatic Habitat Characterization:** In addition to searching for amphibian larvae, the pond/lake nearshore habitat will be characterized at each about 2 m² station. If a site is small, one estimate can be made for the whole site.

    (a) Visualize an about 2 m² box that extends from the shoreline out into the water.

    (b) Estimate and record the percentage of the box's surface area occupied by floating or emergent vegetation.

    (c) Circle one abbreviation for the dominant substrate within the aquatic section of the box.

    (d) Repeat this procedure at each station established for amphibian sampling.

(6) **Captures:** Captured individuals should be identified to species whenever possible. However, do not guess when you are unsure. Instructions for processing animals are provided below. Release identified specimens at the point of capture.

    (a) Species names and codes to be used are as follows:

        i. Each species and life stage of a species detected should be recorded

Code	Scientific Name	Common Name
Amphibians		
ASTR	*Ascaphus truei*	tailed frog
AMGR	*Ambystoma gracile*	northwestern salamander
AMMA	*Ambystoma macrodactylum*	long-toed salamander
BUBO	*Bufo boreas*	western toad
DICO	*Dicamptodon copei*	Cope's giant salamander
DITE	*Dicamptodon tenebrosus*	Pacific giant salamander
ENES	*Ensatina eschscholtzii*	ensatina
PLLA	*Plethodon larselli*	Larch Mountain salamander
PLVA	*Plethodon vandykei*	Van Dyke's salamander
PLVE	*Plethodon vehiculum*	western red-backed salamander
PSRE	*Pseudacris regilla*	Pacific treefrog
RAAU	*Rana aurora*	red-legged frog
RACAS	*Rana cascadae*	Cascade frog
RACAT	*Rana catesbeiana*	bullfrog
RHOL	*Rhyacotriton olympicus*	Olympic torrent salamander
TAGR	*Taricha granulosa*	rough-skinned newt
AMSP	*Ambystoma species*	larval Ambystomatid salamander
Reptiles		
CHBO	*Charina bottae*	rubber boa
ELCO	*Elgaria coerulea*	northern alligator lizard
SCOC	*Sceloporus occidentalis*	northwestern fence lizard
THEL	*Thamnophis elegans*	western terrestrial garter snake
THOR	*Thamnophis ordinoides*	northwestern garter snake
THSI	*Thamnophis sirtalis*	common garter snake
Fish		
BKTR	*Salvelinus fontinalis*	brook trout
BLTR	*Salvelinus confluentus*	bull trout
COTT	*Cottidae*	sculpin
CUTT	*Oncorhynchus clarki*	cutthroat trout
FISH	*unknown fish species*	unknown fish species
KOKA	*Oncorhynchus nerka*	Kokanee
RABO	*Oncorhynchus mykiss*	rainbow trout
SALM	*Salmonidae*	unknown salmon/trout species

on a separate line of page 3 on the datasheet. Record the species name or code, and life stage. For species that lay egg masses, count or estimate the number of masses seen and indicate EM as the life stage on the datasheet. Do not remove individual eggs from masses and do not detach egg masses from supporting vegetation. For species that lay single or small clusters of eggs, count or estimate individual eggs and indicate E as the life stage on the datasheet.

ii. If there is time, measure the snout-vent length (SVL) and total length (TL) of up to 10 individuals of each stage of each species and record on the datasheet.

iii. If it is possible to count the individuals seen, record the count and circle *A* to indicate that this is an *actual* count. If large numbers of individuals are present, record an estimate of the number present and circle *E* on the datasheet to indicate an *estimated* count. The number of animals can be estimated by counting all animals in a smaller section and extrapolating to estimate the total number observed. For example, if a 2 × 1 meter section had a similar density of eggs across the whole area, one could count the number of eggs in several 20 × 20 centimeter boxes, average these values and multiply by 50 (because there are 50-20 × 20 cm boxes in a 2 × 1 m area).

iv. Indicate whether or not the animal detected was calling.

v. Record the method that the amphibians were sampled (H - Hand collected, V - Visual only, or A - Aural only).

(b) **Processing amphibians:** Instructions for processing animals are adapted from Corkran and Thoms, 1996:

i.   **Handling amphibians:**

(1st)   Amphibian skin is very permeable, so it is important that hands are clean (i.e., free of any toxic substances like insect repellent or sunscreen) when handing animals.

(2nd)   Because of this skin permeability, amphibians also desiccate quickly. They should be kept moist and cool at all times, and hands should be wet when handling amphibians directly.

(3rd)   Salamanders are capable of tail autotomy and therefore should not be held by their tails. Tail loss is highly stressful on an animal and should be avoided if at all possible.

(4th)   Tadpoles, hatchlings, and larval salamanders should be placed in individual plastic bags with at least enough native water to cover the entire body. The bags should be kept cool and shaded, and the animals should be processed as quickly as possible. Release the animals at point of capture.

(5th)   Adult salamanders may be held in the hands, but placing them in a plastic bag reduces heat and desiccation stress. Keep bags cool and shaded, process animals as quickly as possible, and release at point of capture.

(6th)   Adult anurans are best examined when the hind legs are fully extended and gently held together, and the body is supported by either hand. Keep the animal moist and cool, process quickly, and release at point of capture.

ii.   **Measuring salamanders:**

(1st)   **Total length (TL):** For hatchling salamanders, total length should be measured instead of snout-to-vent length because of the difficulty in determining where the body ends and the tail begins. The total length is measured from the tip of the snout to the tip of the tail. If recording the measurements for a hatchling salamander, indicate (TL) after the number on the datasheet for clarity.

(2nd)   **Snout-to-vent length (SVL):** Because salamander tails are often shortened due to autotomy or tail bites, snout-to-vent length is the preferred metric for gauging size. The SVL is measured from the tip of the snout to the anterior corner of the vent.

iii.   **Measuring anurans other than tadpoles:**

(1st)   **Snout-to-vent length (SVL):** With the exception of the male tailed frog, anurans do not have tails and are therefore measured from snout to vent. The SVL includes the length of the head and the body, and is measured from the most anterior end to the most posterior end of the body.

iv.   **Measuring anuran tadpoles:**

(1st)   **Snout-to-vent length (SVL):** The SVL for tadpoles includes the length of the head and body but not the tail muscle and fin. If the tadpole is out of water, measure along the dorsum from the tip of the snout to where the body ends. If the tadpole is contained in a plastic bag with water, view the tadpole from underneath and measure along the ventrum.

(7) To conserve time, searches for a particular species should cease once evidence of current breeding activity (e.g., eggs/embryos, tadpoles or metamorphs from the current breeding season) for a species has been recorded.

(8) **Fish Detection:** While characterizing habitat and surveying for amphibians look for any evidence of the presence of fish. Watch for fish in shallows and rising to feed on the surface (but do not be misled by salamanders that surface to breathe and create ripples similar to fish). If fish are not detected in the main body of water, search 30 meters up inlet and outlet streams where fish may be more visible. If fish are detected, indicate the species, if determined, in the species column of the datasheet.

5. **Completing the Datasheet:**

   a. Fill in the remaining spaces on the datasheet with the requested information. If there are any questions about the data to be recorded, see SOP 11 D that defines the fields on the datasheet, or the data sheet definitions tables in Appendix VII.

b. Use the notes section on page 4 of the datasheet to record any additional information of interest. Crew members are strongly encouraged to write notes about features of each site that are not captured by the required data fields. Notes also should explain data values that appear unusual, but are actually correct. The notes will help those who were not present during the survey better understand the data that were collected, the conditions at that site, how difficult it was to get to a site, etc.

6. **Data Management:**

   a. Prior to leaving a site, one crew member (other than the one who recorded the data) will check over the datasheet to make sure that all data fields have been filled (with data or NA), that the recorded data are legible, and that the recorded data do not have any obvious errors.

   b. After the data have been checked the crew member will initial the Field QA field on the top of page 1 of the datasheet and record the end time for the visit.

SOP 10. Estimating PAO for Aquatic Amphibian Surveys
Version 1.00

Revision History Log:

Previous Version Number	Revision Date	Author	Changes Made	Reason for Change	New Version Number

Data Analysis:

To track long-term trends in amphibian populations that occupy the montane lentic habitats of parks, we will apply a multi-season proportion of sites (or area) occupied analysis (PAO) to the amphibian occurrence data collected at the randomly selected ponds/lakes. PAO is an unbiased estimator for the proportion of study sites, in this case ponds or lakes, occupied by a species. This model also incorporates a measurement of detectability of a species, based on the within-season revisits. In addition, it can be used to test the importance of site- and survey-specific variables. Finally, PAO will enable us to estimate the importance of local colonization and extinction processes in the park amphibian populations. This analysis will be completed using the program PRESENCE (MacKenzie and others, 2002, 2003) that is freely available from the USGS website http://www.mbr-pwrc.usgs.gov/software.html listed in the software archive or directly from D. MacKenzie at www.proteus.co.nz.

The first PAO analysis will be completed after the first full panel rotation, 10 years from the beginning of the study. At that point in time, the 50 randomly selected individual sites will have been visited twice within a season during two seasons separated by 5 years. The following section outlines the steps that need to be taken to complete a multi-season PAO analysis. However, the reader also should be aware that PRESENCE has a very useful help file that may aid in both data formatting and analysis.

A. Data need to be arranged in a spreadsheet format such as MS Excel. Make a table where each row represents a separate site. Each site visit is placed in a column (Visit1, Visit2, etc.). The detection data are represented by "1" (species detected) and "0" (species not detected) at each visit. Populate the cells of the table with 0 and 1 unless a site was not visited. In this case, enter a dash (-). These three values (0,1, -) are the only inputs that will be recognized by the software. Open PRESENCE and cut and paste the data table (not including site names or other data besides 0,1, -) into the PRESENCE detection data table.

B. Covariate data can be entered in two categories: site-specific and sampling-occasion. Site-specific covariates do not change throughout the study and include such variables as elevation, habitat type, and site coordinates. These are most likely to affect species presence at a site. Sampling-occasion covariates, like water temperature or pH, are more likely to affect detectability.

C. Build a table for each covariate in the same manner as the detection data. If the covariate data are continuous (e.g., elevation), the actual values can be entered. For categorical covariates, use an indicator value. For example, if a site is characterized as man-made or natural in origin, use 0 to indicate one state and 1 to indicate the other. If there are three categories (i.e., man-made, natural, altered), use two indicator variables (A1 and A2) such that man-made could be represented by $A1 = 1$ and $A2 = 1$, altered is represented by $A1 = 1$ and $A2 = 0$, and natural is represented by $A1 = 0$ and $A2 = 0$.

D. Once detection and covariate data have been entered, select the multi-season PAO model from the analysis toolbar to estimate site occupancy and detection estimates for each species of interest. Indicate the number of seasons (years), sites, and sampling occasions. The secondary sampling periods for our study will be two for each primary sampling period (season denoted by t). Select the model parameterization desired for the analysis. For monitoring efforts, we will be most interested in the models where PAO will be estimated for each season as well as either colonization or extinction. Overall trend analysis, rate of change in occupancy, denoted by λ, will be completed as well. Once a model parameterization has been chosen, the model can be defined using the design matrix that enables the researcher to incorporate site and sampling covariate information. Varying the parameters, including covariates, involved in each candidate model will allow us to test specific hypotheses, such as whether drought or pond permanence might effect amphibian population dynamics in the park.

E. The model that best explains the observed data will be selected from the group of potentially suitable models using an information-theoretic method, Akaike's Information Criterion (AIC) (Burnham and Anderson, 2002). PRESENCE calculates AIC for each model as part of the multi-season analysis. The model with the smallest AIC value is most parsimonious and best fits the observed data. If there are similarly weighted models, a process of model averaging will be applied to account for uncertainty in best model selection (Burnham and Anderson, 2002).

F. It should be noted that PAO also can be calculated for stations within each site if this metric is of interest to resource managers and (or) researchers.

SOP 11. Use of Personal Digital Assistants for Aquatic Amphibian Visual Encounter Survey Data Collection
Version 1.00

Revision History Log:

Previous Version Number	Revision Date	Author	Changes Made	Reason for Change	New Version Number

This SOP gives step-by-step instructions for recording data from aquatic amphibian visual encounter surveys using personal digital assistants (PDAs) and Pendragon Forms©. This method is an alternative to paper data collection and manual data entry. Personal digital assistants use survey-specific nested forms that are pre-loaded with data options to maximize efficiency and minimize error. The data collected on PDAs are identical to those collected on paper, but the arrangement of the data fields is different due to the linking of the forms and the electronic downloading of the data into a database. This SOP provides basic instructions on how to navigate through the forms for an aquatic amphibian visual encounter survey. Refer to section D of this SOP for detailed descriptions of data field definitions.

A. **Getting Started:**

1. Turn on PDA using the power button.

2. Select iForms from the main application window. A single form "SurveyData" will appear on the screen. From here you have two options: "New" or "Review." The site names have been pre-entered on the PDA. To access these site names, tap "Review" and choose the site. If you arrive at a site that is not on the list, return to the "SurveyData" screen and tap "New". From here you can create a new site record.

a. **For a new site:** After tapping "New", a new SurveyData record will appear. Populate the data fields as appropriate. Once selected, many fields are programmed to pull up lists (called "lookup lists") from which an entry is chosen. If you tap a field and no lookup list appears, enter an appropriate value using the keyboard and/or number pad. The keyboard is accessed by tapping the "abc" icon at the bottom left of the writing area on the PDA. The number pad is located at the bottom right of the writing area and is accessed by tapping the "123" icon.

b. **For an existing site:** After tapping "Review," select the appropriate site from the list of site records. Some of the fields will already be populated. Amend these fields if necessary (e.g., add habitat notes, change UserID, etc.)

B. **Navigating the Forms:**

1. **In general:**

a. The PDA is set to automatically turn off after being idle for three minutes (to conserve battery power). Pressing the power key will turn the PDA back on and return you to the same place you were when the power went off.

b. Forms are set up to prohibit deletions once data fields are populated. If you start a record and want to delete it, just write "delete" in the notes/comments section at the end and the data manager will deal with the deletion once you have returned from the field. There are some instances when the PDA will give you the option of deleting a record, in which case proceed as you see fit. This usually happens when you accidentally key "New", generate a new record, and then try to exit without entering any data.

c. Learn to recognize icons and how they change. A single subform field, for instance, looks like a blank sheet of paper with the bottom right corner folded up. When data are entered in the subform, the icon then looks like a sheet of paper with lines on it. Subform lists look like a pair of pages, one filled and one blank.

d. Do not leave fields blank for zero values. For example, if there is no emergent vegetation, then enter "0" instead of leaving it blank. This way it is clear that you did not just forget to collect the data. Conversely, do not enter "0" in fields for which you are unable to collect data. For instance, if the thermometer is not working, do not enter "0"…otherwise it could be mistaken for a valid temperature.

e. Modify existing data by tapping the appropriate field and either entering the correct value or, if the field is a lookup list, by selecting another value.

f. There are at least three versions of lists that you may encounter using the PDA: popup lists, lookup lists, and multi-selection lists. Popup lists are short and cannot be edited or modified. Lookup lists tend to be long, and may be edited and/or modified (depending on how they were formatted). Multi-selection lists are of varying length and cannot be edited or modified. They differ from popup and lookup lists in that multiple values can be selected simultaneously. If the occasion arises where a necessary value is not included in a pop-up or multi-selection list, use the notes/comments section at the bottom of the form to describe the value and alert your supervisor as soon as possible.

g. **Record view vs. Field view:** There are two ways to view fields on the PDA. "Record view" is a two-column format that allows the user to view up to 11 fields on the screen at one time. "Field view" shows only one field per screen, and users scroll through the fields one at a time. The forms default to "Record view", but there are some instances in which the field will be in "Field view" for you to enter data (e.g., a multi-selection list). When you have finished entering data in 'Field view', tap "Record view" to return to the form. Tapping "End" will cause you to exit the record and return you to the record list.

h. The speed with which the PDA can accept new data slows down considerably as more and more data are stored. Be patient and be deliberate while keying in data…sometimes tapping too fast or too hard can cause the PDA to quickly skip back a form or two.

2. **In Detail:**

a. As you create records within a subform, a list of these records is generated. Once you complete a record and tap "End", you will be returned to a list of records within the parent form. Tapping "Done" will return you to the parent form.

b. Coordinates should be recorded in decimal degrees, WGS84 datum.

c. Each station within a site requires a new "Habitat (Station)" record. Station names should be pre-loaded into the PDA. If there are unaccounted-for or un-named stations, create a new one.

d. On long lists, like "County" or "Species", key in the first letter of the entry you are trying to find in the small lookup field at the bottom of the screen. For example: If you are in the "County" field and want to find "Whatcom", type "W" into the lookup field and the list will automatically jump to the W's.

e. Once you complete a record and tap "End", you will be returned to a list of records within the parent form. Tapping "Done" will return you to the parent form.

C. **Data Management:**

1. **QA/QC:** At the end of every survey, the person recording the data will give the PDA to another crew member for review. The second person should review all records from the current site for completeness and accuracy. After each form is reviewed, the reviewer checks the "Verified" or "Checked" box at the end of each form. After QA/QC is completed, back up the data to the backup module.

2. **Backup protocol:** The PDA should be backed up at the end of each survey after it has been checked by a second person and before moving on to the next site. At the end of the day, all data should be backed up on an additional module not carried with the rest of the field gear. If anything happens to the module in the PDA or the PDA itself, then only one day's worth of data are lost.

D. Forms and Subforms:
This is a chart outlining the forms setup:

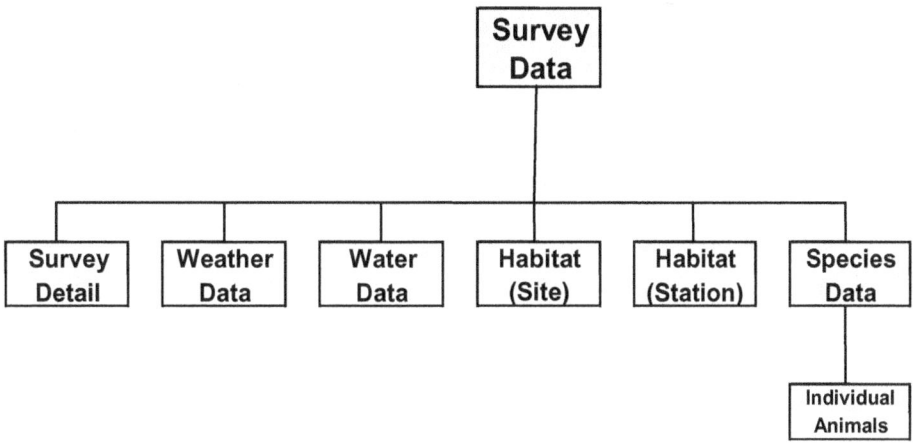

1. **Survey Data**
Tap "Review" to view existing site records. If you are at a site whose data have not been pre-loaded, tap "Add" to add a new record.

Field	Definition
SurveyDate	Select the date and time of the survey.
SiteName	This field displays the name of the survey site.
VisitNum	Key in visit number for the site. This field is required, and you will not be able to leave the form until you enter a value. If you are unclear about the visit number, enter "99" and note it in SurNotes (below).
SurveyDetails	This is a list of subforms through which you navigate during the survey. *SurveyDetails* should be completed first, with the remainder (*HabitatDataSite, WeatherData, WaterData, HabitatDataStation*, and *SpeciesData*) to be completed in any convenient order.
SurNotes	Enter notes if appropriate.
UserID	Name of the surveyor recording data.
Verified	Checkbox to be completed by the person doing QA/QC in the field.

2. **Survey Detail**

Field	Definition
TBegin	The date and start time of the survey
TEnd	The date and end time of the survey
TTotal	This field automatically calculates the time spent (in minutes) conducting the survey.
NumObservers	Number of people conducting survey
Observer1	Lookup list – populate as appropriate
Observer2	Lookup list – populate as appropriate
Observer3	Lookup list – populate as appropriate
Observer4	Lookup list – populate as appropriate
Notes	Enter notes if appropriate.
UserID	Name of the surveyor recording data.
Checked	Checkbox to be completed by the person doing QA/QC in the field.

3. Habitat Data (site)

Field	Definition
SiteName	This field auto-populates based on the selection at the beginning of the survey.
PointName	Choose "site" as the point name for site-level habitat data.
Elevation	Enter elevation in meters
WaterPresent	Checkbox indicating whether or not there is water at the survey site
FishPresent	Checkbox indicating whether or not there are fish present at the survey site.
WaterSource	Indicate the origin of the survey site ("Site Origin" on datasheets).
Permanency	Indicate the permanency of the survey site.
BeaverCode	Indicate any evidence of beaver detected at the survey site.
AvgLength	Average length of aquatic habitat surveyed (in meters).
AvgWidth	Average width of aquatic habitat surveyed (in meters).
PctShallows	Indicate the percent of the total survey site that is < 0.5 m deep.
MaxDepthEst	Estimate the maximum depth of the survey site in meters, and choose either "<1", "1-2", or ">2".
PctAreaSearched	Estimate percentage of entire site searched.
PctEmergVeg	Estimate the percent cover (to the nearest 5 percent) of emergent/floating vegetation for the entire site.
HabitatNotes	Enter notes as appropriate.
UserID	Name of the surveyor recording data.
Checked	Checkbox to be completed by the person doing QA/QC in the field.

4. Weather Data

Field	Definition
SiteName	This field auto-populates based on the selection at the beginning of the survey.
PointName	Choose "site" as the point name for weather data.
StartAirTemp	Air temperature at beginning of survey.
StartWaterTemp	Water temperature at beginning of survey.
TempUnits	Defaults to "C", change if necessary.
StartWindSpeed	Choose the wind conditions present at the beginning of the survey.
StartSky	Choose the weather conditions present at the beginning of the survey.
WeatherNotes	Enter notes if appropriate.
UserID	Name of the surveyor recording data.
Checked	Checkbox to be completed by the person doing QA/QC in the field.

5. Water Data

Field	Definition
SiteName	This field auto-populates based on the selection at the beginning of the survey.
PointName	Choose "site" as the point name for water data.
WaterSample1	Indicate what type of water sample was taken, if any.
WaterSample2	Indicate second type of water sample taken, if any.
SampleLabel1	Exact text of label attached to WaterSample1.
SampleLabel2	Exact text of label attached to additional water sample (indicate which sample).
SampleLabel3	Exact text of label attached to additional water sample (indicate which sample).
WaterColor	Enter whether water is "clear" or "stained."
Transparency	Enter whether water is "clear" or "opaque."
TranspMeth	Enter the method by which you determined water transparency (defaults to "Visual Estimate").
WaterNotes	Enter notes if appropriate.
UserID	Name of the surveyor recording data.
Checked	Checkbox to be completed by the person doing QA/QC in the field.

6. Habitat Data (station)

Field	Definition
SiteName	This field auto-populates based on the selection at the beginning of the survey.
PointName	Choose consecutive letters for each station's point name. The number of stations is determined by the size of the survey site, so you may have to either add new letters or not use all of the ones available.
PctEmergVeg	Estimate the percent cover (to nearest 10 percent) of emergent/floating vegetation at the station.
DominantSubstrate	Dominant substrate in the aquatic part of station (choose from list).
Notes	Enter notes as appropriate.
UserID	Name of the surveyor recording data.
Checked	Checkbox to be completed by the person doing QA/QC in the field.

7.   **Species Data**

Field	Definition
SiteName	This field auto-populates based on the selection at the beginning of the survey.
PointName	Choose "site" as the point name for species data.
Genus	Choose Genus from lookup list. "Species" and "CommonName" will autopopulate.
Species	Choose Species from lookup list. "Genus" and "CommonName" will autopopulate.
CommonName	This field autopopulates when either the "Genus" or "Species" fields are filled.
SCount1	Total number of animals captured within species and age class.
SpeciesAge	Age class of species for which record was created (required field).
Calling	Indicate whether or not species was calling.
Detection	Indicate how the species was detected (required field).
CountMethod	Indicate whether the species count was estimated or actual.
OtherTaxa	List any other taxa present at survey site not otherwise accounted for.
Voucher	Indicate what kind of voucher was taken, if any.
VoucherNum	Indicate the number assigned to the voucher specimen.
IndivAnimals	A single subform. Tapping this will automatically take you to a list of IndivAnimals records.
SpeciesNotes	Enter any notes as appropriate.
UserID	Name of the surveyor recording data.
Checked	Checkbox to be completed by the person doing QA/QC in the field.

8.   **Individual Animals**

Field	Definition
PointName	Field autopopulates with the point name chosen in SpeciesData.
Gender	Gender of animal captured.
SVL	Snout-vent length (mm) of individual.
TL	Total length (mm) of individual.
IANotes	Enter any other notes regarding individual animal.
UserID	Name of the surveyor recording data.
Checked	Checkbox to be completed by the person doing QA/QC in the field.

SOP 12. Fish Sampling and Surveys
Version 1.00

Revision History Log:

Previous Version Number	Revision Date	Author	Changes Made	Reason for Change	New Version Number

Salmonids (*Oncorhynchus sp.*, *Salmo sp.*, and *Salvelinus sp.*) are the most common fish encountered in mountain lakes of the Pacific Northwest, although salmonids are not native to most of these systems. Mountain lakes were originally stocked with fish by private individuals and/or various fisheries agencies (Bahls, 1992). Stocking records typically provide information regarding species presence, relative abundances from stocking densities, and approximate ages of fish present. Information available from stocking records is valuable preliminary data, but is limited for other uses (e.g., individual fish condition, growth measurement, and population recruitment). Often stocking documents are inaccurate or infrequently updated. Accuracy of data originating from such documents should be confirmed by survey and sampling.

A. **Estimating Fish Population Size using Mark-Recapture Methods (Gresswell and others, 1997):** The mark-recapture method described assumes sites contain closed populations and the ability to trap fish is equal between marking and recapture efforts.

   1. **Initial Capture and Marking of Fish:**

      a. Fish are captured using rod and reel. Barbless hooks are used to reduce injury to fish, fish handling time, and to increase fish survival.

      b. Captured fish should be handled with great care and kept moist during processing.

      c. Immediately upon capture, fish should be placed into a collapsible 5 gal plastic container filled with lake water.

      d. After a fish is caught and is ready for processing, it can be placed into a water-filled, relatively small rectangular plastic container. At this time total and fork lengths are measured (in millimeters), the fish is weighed (in grams), and scissors are used to mark the fish by removing the adipose fin or a portion of another fin. This information should be recorded on the appropriate data sheet.

      e. Fish that bleed from the mouth or gills following hook extraction should not be marked or included in final tabulations. The exclusion of injured fish from calculations improves the accuracy of models that estimate fish population size.

      f. Non-injured fish should be released back into the site immediately after processing, although they should be held in the shallow nearshore area until the crew member thinks the fish are ready for complete release.

      g. Gill nets also can be used to capture fish for marking, although their use should be concordant with low water temperature and the nets need to be constantly tended by crew members.

      h. Fish population models that use mark-recapture techniques become more reliable with a greater number of marked fish in a population. Thus, rod and reel angling effort to generate marked fish should be at least 2 days for small lakes (surface area ≤ 5 ha) and 3 days for larger lakes (> 5 ha), depending upon the number of marked fish generated.

i. However, standardization of effort to generate marked fish warrants further discussion as the reliability to estimate fish population size is dependent upon the number of marks generated in a population, and fish population size is not necessarily dependent upon surface area of mountain lakes.

2. **Recapture of Marked Fish:**

   a. Monofilament gill nets with variable-sized mesh can be used to attempt the recapture of marked fish.

   b. This effort is typically conducted no less than 12 hours and preferably 24 hours following the fish marking effort.

   c. Gill nets, set at several locations to maximize the recapture effort, should be set in a lake with one end of each net affixed to shore and the other end extending out toward the center of the lake. A weight or rock-filled bag should be attached to the lead-line of the net end extending out into the lake, and a 250 mL Nalgene® bottle half-filled with water should be attached to the float-line of this end of the net.

   d. Crew members should select gill net fishing locations that will provide optimum fish capture and minimum damage to nets. Ideal locations for setting gill nets include deeper water areas and areas lacking excessive submerged objects (e.g., large boulders, large woody debris, etc.).

   e. During daylight hours, captured fish should be removed from each gill net periodically to improve net capture efficiency.

   f. Setting gill nets overnight increases the effectiveness of the nets.

3. **Gill Net Deployment and Removal and Processing of Fish:**

   a. To deploy gill nets, a crew member holds the majority of the net on shore while one or two crew members in an inflatable boat pull the net end with the bottle float and stuff sack anchor toward the center of the site.

   b. When the entire length of the gill net is pulled from shore to the center of the site, the float and anchor end of the net is released, and the net is allowed to sink to the bottom of the lake. *Note: In lakes with large amounts of coarse woody debris on the lake bottom, the gill net(s) can be allowed to float rather than sink. This will potentially decrease the likelihood of net snagging.*

   c. The crew member on shore secures the shore end of the gill net to vegetation or a rock cairn at the immediate water-land interface. The shore end of the net is positioned at the water-land interface to reduce incidental capture of birds or mammals.

   d. Crew members in the boat should visually inspect the gill net for tangles, snared woody debris, or other anomalies that may affect the ability of the gill net to capture fish.

   e. To remove fish from gill nets, two crew members work in unison from the inflatable boat. Starting from the shore end of the net, one crew member holds the float line of the net while the other crew member removes the entangled fish.

   f. Care is taken to prevent damage to the net while removing fish. Captured fish are placed into a plastic container or garbage bag until all captured fish are removed from the net. Captured fish are then taken to shore for processing.

   g. If more fish are required for population size estimates, each net is reset and allowed to fish until a suitable number of fish are captured for population model calculations.

   h. The processing of captured fish includes: (1) species identification; (2) identification of recapture status (i.e., marked and unmarked; (3) measurement of total length, fork length, and the weight of each fish; (4) collection of scale and otolith samples; (5) determination of sex; and (6) stomach contents of selected individuals.

   i. Scale and otolith samples can be removed from individual fish and stored in small coin envelopes or centrifuge tubes for future processing.

   j. Stomach content samples should be stored in plastic Ziploc® or WHIRL-PAK® bags or HDPE bottles and preserved in 70 percent ethanol.

   k. All data should be recorded on the appropriate data sheet.

   l. After the collection of meristic data and samples has been completed, carcasses should be properly disposed. Depending on the requirements of the agency involved in the monitoring program, carcasses can be: (1) buried at least 200 m from the site; (2) deposited in the site over its deepest point after each fish's swim bladder has been punctured; or (3) packed out of the field in plastic garbage bags for disposal in a landfill.

   m. Proper disposal of fish carcasses and washing of equipment used to sample fish reduces the probability of carnivore encounters and the introduction of biological contaminants between sampling sites.

B. **Non-Mark-Recapture Fishing Effort:**

1. When multiple day sampling efforts at a monitoring site are not possible or fish population size estimates are not necessary, either rod and reel angling or gill nets can be used to determine fish presence or absence and species, as well as to collect meristic data (e.g., total and fork lengths, weight, sex, and age).

2. Rod and reel angling is biased toward larger-sized fish, but provides a non-lethal method for sampling fish, especially when barbless hooks are used, when lethal sampling of a fish population is unacceptable.

3. The use of gill nets may provide a better representation of overall fish population than rod and reel angling. Gill net fishing is primarily a lethal sampling method. However, when nets are continuously tended by crew members, they can be a non-lethal sampling method

4. Monofilament gill nets with variable-sized mesh panels catch all sizes (i.e., small to large) of fish.

5. When gill nets are used during single day sampling visits, nets should be set as outlined in SOP 12 A 3 a-d, and should remain in the lake for a minimum of 4 hours.

C. **Cleaning Fish Capture and Sampling Equipment:**

1. In the field, all equipment used to capture and process fish should be cleaned when sampling is completed.

2. Equipment used to collect meristic data (e.g., ruler or measuring board, knife, forceps, weighing scale, holding and processing containers, etc.) should be thoroughly cleaned using water from the site.

3. Before cleaning with water, meristic data collection equipment can be swabbed with Pinesol®, an effective solution for neutralizing the smell of fish and for cleaning fishing gear (David Donald, Ecological Research Division, Environment Canada, oral commun., 2001). If Pinesol® is used, the treated equipment will need to be profusely rinsed with water.

4. The cleaning station should be setup well away from the shore of the study site so that contaminants from the cleaning process do not enter the site.

5. Gill nets should be cleaned of any organic debris (e.g., twigs, branches, aquatic vegetation, moss, etc.). The monofilament nets can also be dipped in Pinesol® and rinsed profusely with water. However, since the amount of Pinesol® required to clean nets in the field may be prohibitive, nets will at least need to be thoroughly rinsed with water and air-dried (if possible) before being stowed for transport from the study site.

6. In the laboratory, all fish capture and sampling gear should be soaked for 24 hours in Pinesol®, soaked for 24 hours in deionized water, then rinsed thoroughly and profusely with deionized water.

7. The equipment then should be allowed to air-dry before being stowed for transport.

8. Periodically during the sampling season and at the end of the season, gill nets should be examined for tears and other damage and mended as required. Instructions for net maintenance can be found in Nielsen and Johnson (1983).

D. **Sampling for Persistent Organic Pollutants (provided by Kim Anderson, Oregon State University):**

1. Collect fish by hook and line.

2. Map exact fishing locations on representational map of site at the time of collection.

3. Kill fish by quick blow to the head.

4. From the time of collection to the time of wrapping, keep fish as clean as possible, and *wrap as soon as possible after catching*. Keep fish away from any contaminants, especially boat fuel and oil, etc. If a fish falls or is somehow contaminated with fuel do not submit this fish for analysis.

5. *Double wrap* each individual fish in aluminum foil, *dull-side next to fish*.

6. Place wrapped fish in a Ziploc® bag (1 fish per bag).

7. Uniquely label each Ziploc® bag. Include on each bag:

    a. Site Identifier

    b. Collection Date

    c. Species

    d. Unique Specimen Identification Number

8. Immediately place fish in cooler with cold packs or keep samples as cool as possible until they can be transported out of the field.

9. If specimens cannot be transported immediately to laboratory for storage and processing, they should be stored at 4°C until they can be transported to the processing laboratory. Further, specimens should continue to be stored at 4°C until sample preparation.

10. **One final note: It is extremely important that these samples be kept clean and uncontaminated. Samples should be handled with utmost care.**

SOP 13. Data Management
Version 1.00

Revision History Log:

Previous Version Number	Revision Date	Author	Changes Made	Reason for Change	New Version Number

Meticulous data management is essential to the success of any monitoring project. The proper collection, recording, analysis, and archiving of data should be a top priority of each project crew member.

A. Data Collection and Management in the Field:

1. Prior to leaving for the field, crew members should confirm that all data sheets to be used to record sample collection and measurement information have been packed along with all sampling equipment and gear (see Appendix I).

2. Each sampling SOP will require the recording of general as well as SOP specific information. A data sheet used to record sample collection and measurement information can be either for a single SOP or used to record information related to two or several sample collection SOPs.

3. General information to be recorded on each field data sheet includes: A) site identifier; B) site date; C) name(s) of crew member(s) responsible for collecting sample or making measurement; D) name(s) of crew member(s) responsible for recording data; E) time of sample collection or measurement; F) weather conditions at time of sampling or measurement; and G) sample specific number if applicable.

4. Data sheets should be constructed so they are not confusing to use and include information entry prompts for the recording of all samples to be collected or measurements to be made and entered on the data sheet.

5. It is essential that the labeling of all samples and/or containers holding samples be non-ambiguous and clearly identified on appropriate data sheets.

6. A digital camera should be used to take photographs at each monitoring site. However, if photographs are to be taken using an analog camera, then film roll and picture sequence numbers should be clearly and accurately recorded for each site. This information should be recorded in the monitoring project field book.

7. Sample collection information (e.g., sample type, number, date, and time) should be entered into the monitoring project field book at time of sample collection or measurement.

8. Prior to leaving each site, the crew leader should check over the field data sheets to confirm that all fields (information entry prompts) have been filled-in (i.e., data or NA) on each sheet, that all recorded information is clearly legible, and that the reported information does not have any obvious errors.

9. The crew leader or another crew member also should be responsible for confirming that sample identification information on sample container labels are consistent with sample identification information on data sheets and, if applicable, in the monitoring project field book.

10. Each data sheet should have a check-off line or box labeled "Field Review". Upon completion of review of each data sheet, the crew leader should initial this line or box. Also, a note should be entered into the monitoring project field book that all sample identification information has been checked and confirmed as correct.

11. Upon arriving at and prior to leaving a site, the date and time should be entered into the monitoring project field book

B. Data Management in the Laboratory:

1. Data sheets should be photocopied and then the original and copied sheets should be collected into separate file folders or binders designated for this purpose. This should be done no more than 24 hours after returning from the field. The original and copied data sheet folders or binders should be stored in separate buildings and preferably in fire-safe storage containers.

2. Information from the data sheets and data from samples that require further processing (e.g., concentrations of water chemistry parameters, chlorophyll-*a* concentrations, etc.) should be entered into the appropriate data file (e.g., an Excel worksheet or Access data table) as soon as possible after returning from the field. One person should be responsible for data entry and a second person should be responsible for checking the data entered for any entry errors.

3. A log of data entry and entry quality control checks should be kept. This log should record the following information: **A**) data file name; **B**) sample type; **C**) number of folder or binder containing field or laboratory data sheet ; **D**) names of personnel responsible for data entry and entry quality control; **E**) date of data entry; **F**) date of entry quality control check.

4. Data files should be stored in multiple places, preferably on at least one computer hard-drive, and on at least one external source (e.g., 3.5" floppy disk or CD).

5. Versions of data files and/or the project database should be updated at least once a year. Version designation should follow the following format: Version 1.00 followed by Version 1.01 followed by Version 1.02, etc.

6. Data entry is an essential part of any monitoring project, and time for the processing and entry of data should be built into the work schedule of the project field crew or laboratory support crew.

7. Hiring a data manager as part of the monitoring project staffing would be most desirable.

C. Data Analysis:

1. Data should be analyzed at least annually using a statistical software chosen by the agency responsible for the monitoring project.

2. Analyzed data should be reported at least once a year as part of an annual report describing field sample collection activities and outcomes associated with the analysis of data generated due to the efforts of the monitoring project.

3. Data analysis also can be helpful in reviewing and assessing the importance of parameters being measured as well as revising monitoring project sample collection objectives, regimes, and SOPs.

4. The review of project objectives, sample collection regimes, and SOPs, and parameters being measured and monitored should be undertaken at least before the beginning of each field season. This review will be helpful in making sample collection effort more focused and efficient.

D. Metadata:

Creation of a metadata file is an integral part of any project that collects samples that generate data, data files, and a database. Metadata consists of information that documents and characterizes information contained within data files and databases. This documentation and characterization can be extremely helpful for individuals interested in understanding and/or using the information contained within data files and databases with which they are not familiar. Metadata also creates a readily accessible pool of institutional memory concerning data collected over long periods of time. The process of developing metadata does not result in data archival. However, it does enhance the life of information by providing a standardized description that can be critical in information retrieval.

A metadata file should identify who is responsible for the data, what the data represent, when the data were collected or generated, where the data were collected, as well as why and how the data were collected. Each monitoring project should create a metadata file that conforms to the Content Standard for Digital Geospatial Metadata (or FGDC standard). This file should be created after the first project field season and the creation of data files and a database for the project. This file also should be reviewed annually and updated as required.

SOP 14. Mountain Ponds and Lakes Database
Version 1.00
(prepared by Bret Christoe, Data Manager, NCCN MORA)

Revision History Log:

Previous Version Number	Revision Date	Author	Changes Made	Reason for Change	New Version Number

The North Coast and Cascades Network (NCCN) Mountain Ponds and Lakes Database is intended to accommodate mountain pond and lake monitoring protocol needs for surveys conducted in several NCCN parks. Capabilities will include: (1) adding new survey records; (2) editing existing records; (3) searching and viewing records; (4) data validation; (5) identifying outliers; (6) analysis; and (7) creating exports for complex statistical analysis. To support use as a central network database, it is designed with a SQL Server back-end structure and MSAccess front-end interface.

The back-end is deployed on the Mount Rainier National Park (MORA) GIS server under PWRMORAGIS with a security configuration allowing for remote data entry and remote database management, restricted to appropriate users located in various NCCN parks. The initial database model was guided by Phase II of the Natural Resource Database Template whereby primary keys in the Locations and Events tables are built by concatenating project codes, location names, survey dates and times. These primary keys ensure referential integrity as an autonomous database, and when integrated with other databases. Integration with GIS is supported via the GISCode attribute of tbl_Locations and by employing queries of existing GIS database tables including coordinate, watershed, and habitat information.

Figures 1 and 2 summarize the back-end architecture, depicting major components (Locations and Events) and their respective relationships in the underlying database.

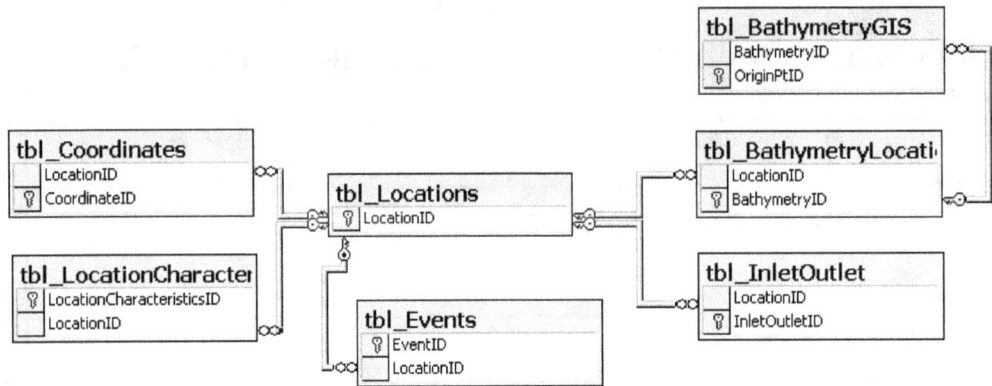

Figure 1. Locations primary and foreign key relationships.

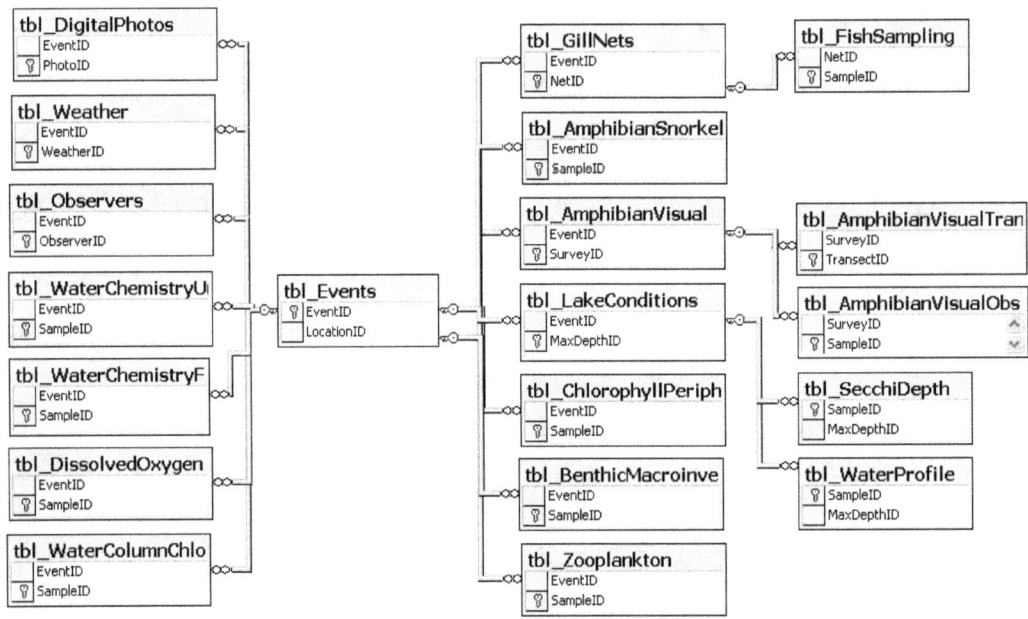

Figure 2. Events primary and foreign key relationships.

Data will be accessed via distributed MSAccess project files (.adp) deployed initially on the MORA GIS Server. For more information contact Bret_Christoe@nps.gov. As the front-end is developed for wider network use, files can be deployed either to a central access point at each park, or as convenient for the individual user (within the NPS firewall). Figure 3 is a walk-through of the front-end application leading up to data entry for a specific survey type (Survey Data forms).

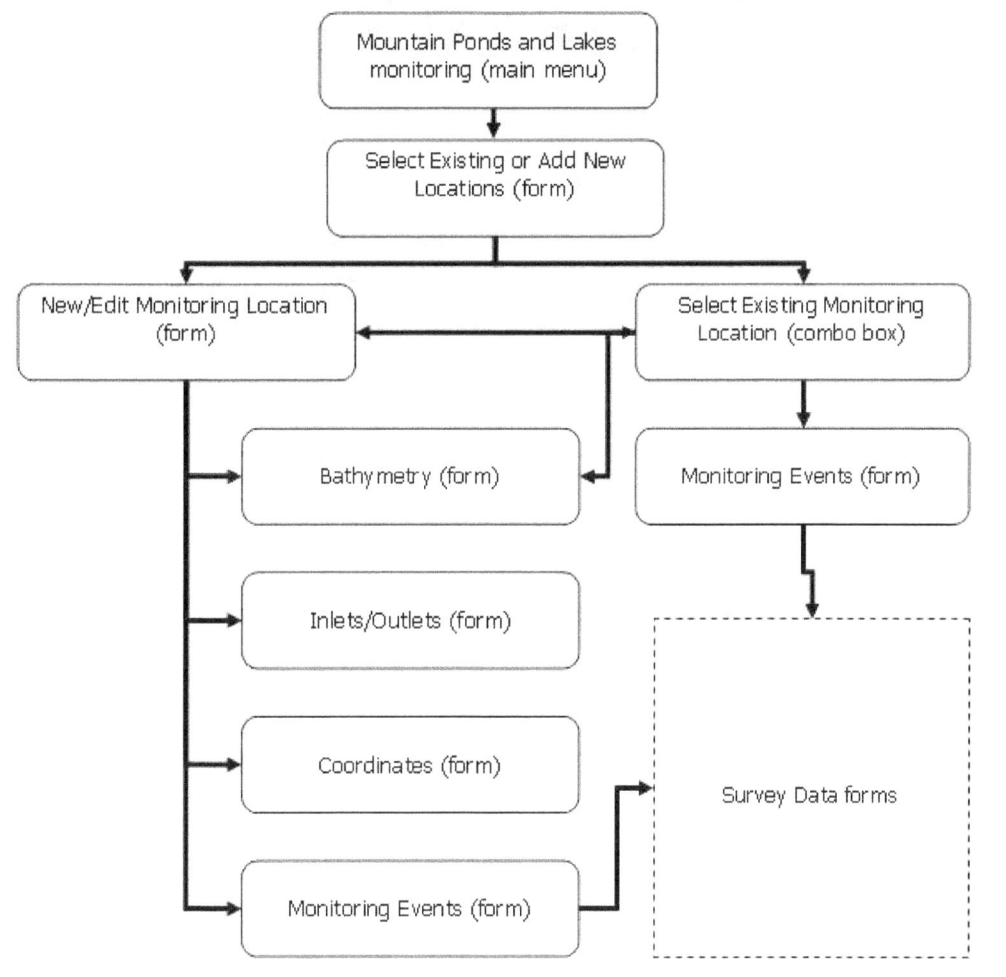

Figure 3.   User interface-flow diagram.

A. Data Entry SOP (example for Amphibian Visual Survey):

1.   On the Main Menu form (figure 4) click the Data Entry button.

Figure 4.   Main menu form, NCCN Mountain Ponds and Lakes Monitoring Database.

2.   From the Locations form (figure 5) select an existing location from the pick list and then click on the Open Monitoring Events for Selected Location button.

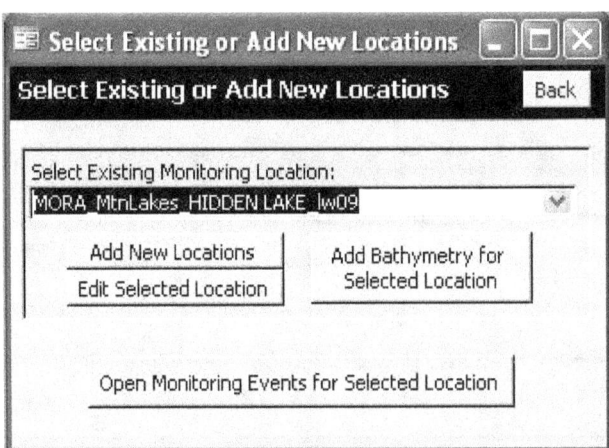

Figure 5.   Locations form, NCCN Mountain Ponds and Lakes Monitoring Database.

3. On the Monitoring Events form (figure 6) click on the New Event button to begin entering event data. The Start Date, Start Time, and Survey Type are required entries. The Event ID, located just below the form header, will be created automatically as the Start Date and Start Time text boxes are completed. Observers, weather data, and digital photos will be entered by selecting among the command buttons along the bottom of the Monitoring Events form. Once all entries are made, select the Biological Observations button to continue.

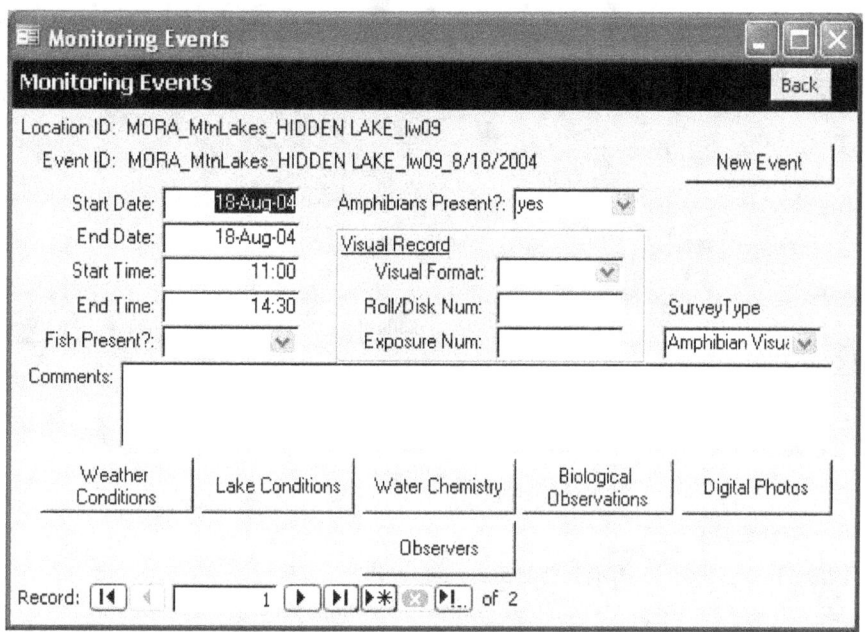

Figure 6. Monitoring Events form, NCCN Mountain Ponds and Lakes Monitoring Database.

4. Selection of the Biological Observations button will cause the Select to Enter Observation form (figure 7) to appear. From the Select a Biological Event option group, select the Amphibians (Visual) check box (be aware that the user will be unable to continue with data entry until a check box is selected). Click the Go to selection button to continue.

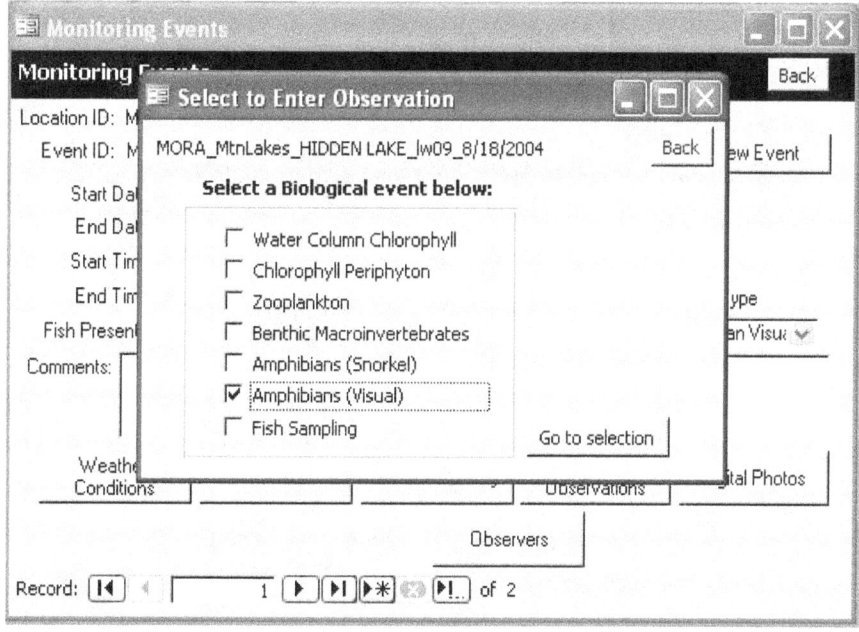

Figure 7. Select to Enter Observation form, NCCN Mountain Ponds and Lakes Monitoring Database.

5.   The Amphibian Visual form (figure 8) and associated sub-forms allow the entry of lake measurements as well as transects and species observations. For a single event, multiple species observations and up to 20 transects can be entered. Begin entering data from the appropriate datasheet. Once entry is complete click on the back button to return to the previous form.

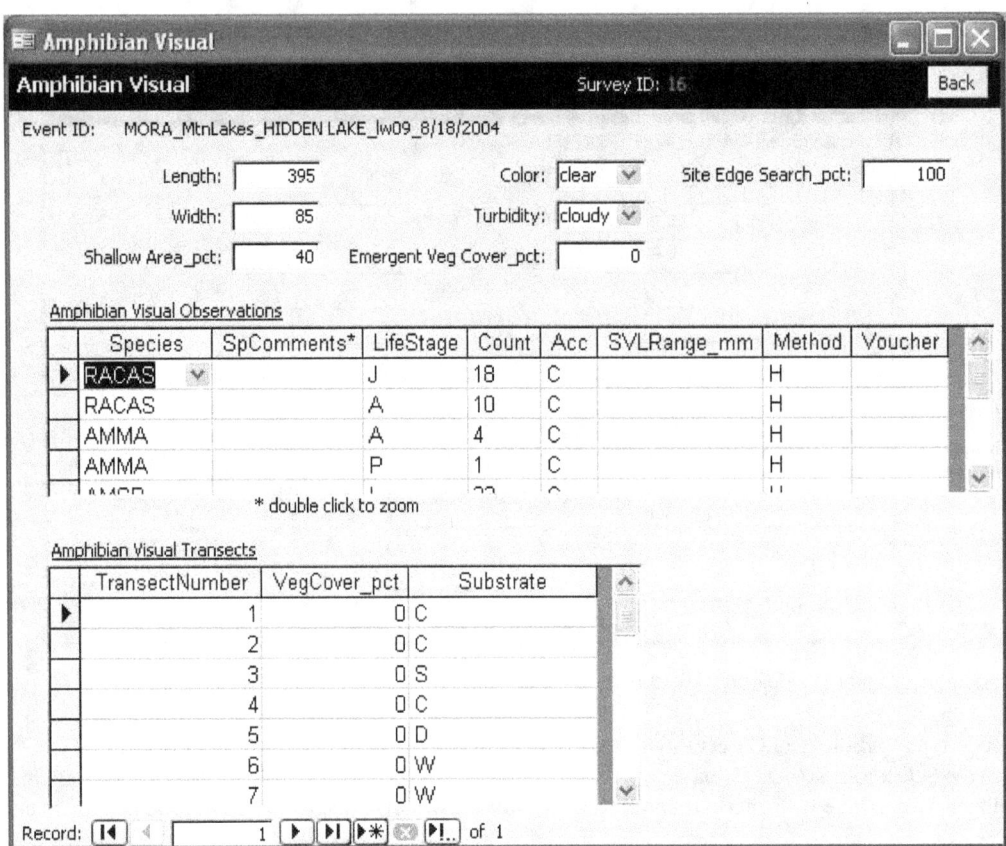

Figure 8.   Amphibian Visual form, NCCN Mountain Ponds and Lakes Monitoring Database.

Comments: This database is in draft form; further field testing may warrant restructuring and revision. Authorized NCCN Park users can access the database and complete documentation by contacting Bret_Christoe@nps.gov**.**

B. Data Dictionary

[**Abbreviations:** A, actual; cm, centimeter; E, estimated; H, hand collected; n, no; ns, nearshore; os, offshore; pk, picked; pr, preserved; y, yes; m, meter; mg/L, milligram per liter; mL, milliliter; μm, micrometer; μS/cm, micorsiemens per centimeter; V, visual observation]

Field Name	Data type	Null	Description
tbl_AmphibianSnorkel - Data table for Amphibian Snorkel survey			
EventID	varchar	No	foreign key to tbl_Events(EventID)
SampleID	int	No	Automated identity, equates to datasheet "sample number"
SurveyType	varchar	YES	amphib snorkel-survey type from datasheet
TransectNumber	int	YES	from datasheet - transect number; 1-20
Species	char	YES	4-5 char species code from qry_tlu_NCCNAmphibians
LHS	varchar	YES	Life History Stage - from tlu_LifeStage
TotalCount	int	YES	species count - from datasheet - count
Habitat	varchar	YES	characterization of habitat - from qry_tlu_Substrate
tbl_AmphibianVisual - Data table for Amphibian Visual survey			
EventID	varchar	No	foreign key to tbl_Events(EventID)
SurveyID	int	No	Automated identity
Length_m	float	YES	survey area length in meters
Width_m	float	YES	survey area width in meters
ShallowArea_pct	smallint	YES	shallow area percentage
Color	char	YES	water color - clear; stained
Turbidity	char	YES	water turbidity - clear; cloudy
EmergentVegCover_pct	smallint	YES	percent emergent vegetative cover
SiteEdgeSearch_pct	smallint	YES	site edge search percentage
tbl_AmphibianVisualObs - Species observations for Amphibian Visual survey			
SurveyID	int	No	foreign key to tbl_AmphibianVisual(SurveyID)
SampleID	int	No	Automated identity
Species	char	YES	species observed - from tlu_AmphibSpecies
SpeciesComments	varchar	YES	comments this species observation
LifeStage	char	YES	from tlu_LifeStage
TotalCount	bigint	YES	total number observed this species
CountAccuracy	char	YES	count accuracy - E;A
SVLRange_mm	char	YES	snout vent length - from tlu_SVLRange
DetectionMethod	char	YES	method used for detection - H;V
VoucherNumber	char	YES	voucher number, if any applied to specimen(s)
EggFungusPresenceYN	char	YES	egg fungus observed - y;n
FishPresenceYN	char	YES	fish observed - y;n
tbl_AmphibianVisualTransect - Transects for Amphibian Visual survey			
SurveyID	int	No	foreign key to tbl_AmphibianVisual(SurveyID)
TransectNumber	smallint	No	Automated identity - index for transect numbers
VegCover_pct	smallint	YES	percent vegetative cover
Substrate	char	YES	lake substrate from tlu_Substrate
tbl_BathymetryGIS - Bathymetry points and depths			
LocationID	varchar	No	refers to tbl_Locations (LocationID)
BathymetryID	int	No	foreign key to tbl_BathymetryLocation
OriginPtID	int	No	automated identity
Origin_East	int	YES	UTM Easting
Origin_North	int	YES	UTM Northing
Distance_m	decimal	YES	tangent in meters
Bearing_deg	decimal	YES	compass bearing in degrees
Depth_m	decimal	YES	depth in meters

B. Data Dictionary—Continued

[**Abbreviations:** A, actual; cm, centimeter; E, estimated; H, hand collected; n, no; ns, nearshore; os, offshore; pk, picked; pr, preserved; y, yes; m, meter; mg/L, milligram per liter; mL, milliliter; μm, micrometer; μS/cm, micorsiemens per centimeter; V, visual observation]

Field Name	Data type	Null	Description
tbl_BathymetryLocation - Bathymetry cross-reference to tbl_Locations			
LocationID	varchar	No	foreign key to tbl_Locations (LocationID)
BathymetryID	int	No	Automated identity
EnteredBy	varchar	YES	automated - windows login name, function is (suser_sname())
tbl_BenthicMacroinvertebrates - BMI survey			
LabID	varchar	YES	for tracking samples sent to lab - LakeCode + Date + MaxDepth
SampleID	int	No	Automated identity
SampleType	char	YES	ns;os
CollectionTime	datetime	YES	time actual collection began
UTME	int	YES	supplemental UTM coordinate, varies from MaxDepth coordinate
UTMN	int	YES	supplemental UTM coordinate, varies from MaxDepth coordinate
HabitatPrimSubstrate	varchar	YES	from qry_tlu_LakesSubstrate
HabitatSecSubstrate	varchar	YES	from qry_tlu_LakesSubstrate
ProcessMethod	varchar	YES	pr;pk
Subsampled_yn	char	YES	y;n
SubsampleAmount	int	YES	whole integer amount of subsample
tbl_ChlorophyllPeriphyton - Chlorophyll Periphyton survey			
EventID	varchar	No	foreign key to tbl_Events (EventID)
LabID	varchar	YES	for tracking samples sent to lab - LakeCode + Date + MaxDepth
SampleID	int	No	Automated identity
CollectionTime	datetime	YES	time actual collection began
UTME	int	YES	supplemental UTM coordinate, varies from MaxDepth coordinate
UTMN	int	YES	supplemental UTM coordinate, varies from MaxDepth coordinate
SampleSubstrate	varchar	YES	from qry_tlu_LakesSubstrate
Chlorophyll	float	YES	chlorophyll concentration, μm
tbl_Coordinates - Coordinates and data for locations in tbl_Locations			
LocationID	varchar	No	foreign key to tbl_Locations (LocationID)
CoordinateID	int	No	Automated identity
UTME	int	YES	UTM Easting - Pick from GIS List or Enter Specific Value
UTMN	int	YES	UTM Northing - Pick from GIS List or Enter Specific Value
Datum	char	YES	reference system used for defining the coordinates
UTMAccuracy	smallint	YES	UTM accuracy from qry_tlu_CoordAccuracy
UTMMethod	varchar	YES	method used from qry_tlu_CoordMethod
Elevation	smallint	YES	elevation in meters
GPSNotes	varchar	YES	note on obtaining coordinates
tbl_DigitalPhotos - Digital Photograph links for various survey types			
EventID	varchar	No	foreign key to tbl_Events (EventID)
PhotoID	int	No	Automated identity
Subject	varchar	YES	photo subject
Purpose	varchar	YES	photo purpose
FileLocation	varchar	YES	hyperlink to file location
tbl_DissolvedOxygen - DO survey data table			
EventID	varchar	No	foreign key to tbl_Events (EventID)
SampleID	int	No	Automated identity
BeginningVolume_ml	decimal	YES	beginning volume in ml
EndingVolume_ml	decimal	YES	ending volume in ml
mgDO	float	YES	dissolved oxygen concentration
DOMethod	varchar	YES	method used

B. Data Dictionary—Continued

[**Abbreviations:** A, actual; cm, centimeter; E, estimated; H, hand collected; n, no; ns, nearshore; os, offshore; pk, picked; pr, preserved; y, yes; m, meter; mg/L, milligram per liter; mL, milliliter; μm, micrometer; μS/cm, micorsiemens per centimeter; V, visual observation]

Field Name	Data type	Null	Description
tbl_Events - Site visit information			
EventID	varchar	No	LocationID + StartDate + StartTime
LocationID	varchar	YES	Foreign key to tbl_Locations (LocationID)
ParkCode	char	YES	the four-character NPS unit code - from qry_tlu_Parks
ProjectCode	varchar	YES	1 - 10 char code assigned to this project
StartDate	datetime	YES	date when sampling began - mm/dd/yy
EndDate	datetime	YES	date when sampling ended - mm/dd/yy
StartTime	datetime	YES	time of day when sampling began - hh:mn of 24 hour clock
EndTime	datetime	YES	time of day when sampling ended - hh:mn of 24 hour clock
TripReport	varchar	YES	hyperlink to trip report
ProtocalVersion	char	YES	version of protocol used
Comments	varchar	YES	comments regarding this sampling event
AmphibiansPresentYN	char	YES	were amphibians present?
FishPresentYN	char	YES	were fish present?
VisualFormat	varchar	YES	from qry_tlu_VisualRecord, format used for image document
Roll_Disk_Num	tinyint	YES	film roll or disk number
ExposureNum	tinyint	YES	film exposure number/ picture number
Subsurvey	varchar	YES	survey type documented - from tlu_EventsSubsurvey
tbl_FishSampling - Fish species observations resulting from gillnet deployments			
NetID	int	No	foreign key to tbl_GillNets (NetID)
SampleID	int	No	Automated identity
FishNum	tinyint	No	fish number from datasheet (not a fish count)
Species	char	YES	from qry_tlu_NCCNFish
Length_cm	tinyint	YES	fork length in centimeters
Weight_g	tinyint	YES	fish weight in grams
Notes	varchar	YES	notes regarding this fish sample
tbl_GillNets - Gillnet deployment data table for conducting fish surveys			
EventID	varchar	YES	foreign key to tbl_Events (EventID)
NetID	int	No	Automated identity
TimeDeployed	datetime	YES	Time net was deployed
TimeRetrieved	datetime	YES	Time net removed from water
Notes	varchar	YES	notes regarding this gillnet deployment
tbl_InletOutlet - Polygon measurements of inlets and outlets			
LocationID	varchar	No	foreign key to tbl_Locations (LocationID)
InletOutletID	int	No	automated identity
UTME	int	YES	UTM Easting
UTMN	int	YES	UTM Northing
InletOrOutlet	varchar	YES	identify inlet or outlet
Description	varchar	YES	describe this feature in plain text
tbl_LakeConditions - Lakes survey data			
EventID	varchar	No	foreign key to tbl_Events (EventID)
MaxDepthID	int	No	automated identity
MaxDepth_m	decimal	YES	Maximum depth in meters
MaxUTME	int	YES	coordinate of maximum depth
MaxUTMN	int	YES	coordinate of maximum depth
SecchiTime	datetime	YES	Time Secchi measurements began

B. Data Dictionary—Continued

[**Abbreviations:** A, actual; cm, centimeter; E, estimated; H, hand collected; n, no; ns, nearshore; os, offshore; pk, picked; pr, preserved; y, yes; m, meter; mg/L, milligram per liter; mL, milliliter; μm, micrometer; μS/cm, micorsiemens per centimeter; V, visual observation]

Field Name	Data type	Null	Description
			tbl_LakeConditions - Lakes survey data—Continued
BottomVisible_yn	char	YES	Was lake bed visible y;n
SecchiCloudCover	varchar	YES	cloud cover for Secchi only
Ripples	char	YES	ripple characterization - none;slight;moderate;heavy
RippleSpread_cm	tinyint	YES	distance in cm between ripples
Glare	char	YES	severity of glare – high;moderate;low;none
ProfileStartTime	datetime	YES	Time vertical (water) profile started
ProfileEndTime	datetime	YES	Time vertical (water) profile ended
ProfileMachine	varchar	YES	Machine used to conduct profile
StratifiedYN	char	YES	is the lake thermally stratified?
			tbl_LocationCharacteristics - Additional location attributes
LocationCharacteristicsID	int	No	automated identity
LocationID	varchar	No	Foreign key to tbl_Locations (LocationID)
Watershed	varchar	YES	qry_tlu_LakesGISWatershed
SurfaceArea_ha	int	YES	surface area in hectares
Perimeter_m	int	YES	shoreline length in meters
VegZone	varchar	YES	from qry_tlu_LakesGISVegZone
DominantVeg	varchar	YES	dominant vegetation - from tlu
BasinAspect	char	YES	orientation of main axis of site - unit of meas. compass direction
WatershedArea_ha	int	YES	watershed area in hectares
BasinMineralComp	varchar	YES	basin mineral substrates
BasinOrigin	varchar	YES	basin geological origin types - include bench, cirque, tarn, ice scour, kettle, moraine, slump, trough, and fault-influenced
MaxDepth	int	YES	maximum depth
SampleFrame	varchar	YES	sampling frame applied
			tbl_Locations - General location data
LocationID	varchar	No	ParkCode + ProjectCode + Lakes(GIS)Code
ParkCode	char	No	the four-character NPS unit code - from qry_tlu_Parks
ProjectCode	varchar	No	1 - 10 char code assigned to this project
LocationName	varchar	YES	describes location
GIScode	varchar	YES	code for associated GIS table - from qry_tlu_LakesGISCode
Established	datetime	YES	date this location established
Discontinued	datetime	YES	date this location discontinued
Shapefile	varchar	YES	location of GIS shapefile
			tbl_Observers - Staff participating in survey event
EventID	varchar	No	foreign key to tbl_Events (EventID)
ObserverID	int	No	Automated identity
ObserverInitials	char	No	observer first, last and middle initials
			tbl_ResultsBMI - BMI survey results
SampleID	int	YES	foreign key from tbl_EventsBenthicMac (SampleID)
Species	varchar	No	species from tlu_NCCNSpecies
			tbl_ResultsZooplankton - Zooplankton survey results
SampleID	int	No	Automated identity
Species	int	No	species from tlu_NCCNSpecies
Number	int	YES	Count of this species

B. Data Dictionary—Continued

[**Abbreviations:** A, actual; cm, centimeter; E, estimated; H, hand collected; n, no; ns, nearshore; os, offshore; pk, picked; pr, preserved; y, yes; m, meter; mg/L, milligram per liter; mL, milliliter; μm, micrometer; μS/cm, micorsiemens per centimeter; V, visual observation]

Field Name	Data type	Null	Description
tbl_SecchiDepth - Secchi disk measurements for water clarity			
SampleID	int	No	Automated identity
MaxDepthID	int	No	Foreign key to tbl_LakeConditions (MaxDepthID)
DescendDepth_m	decimal	YES	depth Secchi disk disappears
AscendDepth_m	decimal	YES	depth Secchi disk reappears
tbl_WaterChemistryFiltered - Store variables measured using filtered water			
EventID	varchar	No	foreign key to tbl_Events (EventID)
LabID	varchar	YES	for tracking samples sent to lab - LakeCode + Date + MaxDepth
SampleID	int	No	Automated identity
Type	varchar	YES	sample type - nutrients; anions/cations; DOC
CollectionTime	datetime	YES	time of collection
Depth_m	decimal	YES	sample depth (m)
Volume_mL	decimal	YES	sample volume (mL)
FilterSize	decimal	YES	filter size (μm)
Comments	varchar	YES	enter comments associated with this sample
tbl_WaterChemistryUnfiltered - Store variables measured using unfiltered water			
EventID	varchar	No	foreign key to tbl_Events (EventID)
SampleID	int	No	Automated identity
CollectionTime	datetime	YES	time of collection
Depth_m	decimal	YES	sample depth (m)
Volume_mL	decimal	YES	sample volume (mL) - at least 250 mL
pH	float	YES	precipitation chemistry
Alkalinity_mgl	float	YES	alkalinity concentration mg/L
Conductivity	float	YES	conductivity μS/cm
AlkGran	float	YES	gran titration
Alkdblendpt	float	YES	double endpoint
tbl_WaterColumnChlorophyll - Water Column Chlorophyll survey data			
EventID	varchar	No	foreign key to tbl_Events (EventID)
SampleFiltered_mL	int	YES	Total volume that was filtered in milliliters
SampleID	int	No	Automated identity
CollectionTime	datetime	YES	time sample collected
Depth_m	decimal	YES	sample depth (m)
Volume_mL	decimal	YES	sample volume (mL)
Chlorophyll	float	YES	chlorophyll concentration
FilterSize	decimal	YES	filter size (μm)
tbl_WaterLevel - Site water-level measurement			
TagNumber	int	No	benchmark tag number - 3 separate measurements
MaxDepthID	int	No	foreign key to tbl_LakeConditions (MaxDepthID)
WaterLevel_cm	int	YES	water level in centimeters
CordDistance_cm	int	YES	distance from cord to surface - centimeters
tbl_WaterProfile - Water Profile data table			
SampleID	int	No	automated identity
MaxDepthID	int	No	foreign key to tbl_LakeConditions (MaxDepthID)
Depth_m	decimal	YES	sample depth (m)
Temperature_c	decimal	YES	sample volume (mL)
DissolvedOxygen_mgl	decimal	YES	dissolved oxygen (mg/L)
pH	decimal	YES	precipitation chemistry
SpecificConductance	decimal	YES	specific conductance (μmhos)

B. Data Dictionary—Continued

[**Abbreviations:** A, actual; cm, centimeter; E, estimated; H, hand collected; n, no; ns, nearshore; os, offshore; pk, picked; pr, preserved; y, yes; m, meter; mg/L, milligram per liter; mL, milliliter; μm, micrometer; μS/cm, micorsiemens per centimeter; V, visual observation]

Field Name	Data type	Null	Description
			tbl_Weather - Weather observations associated with various survey types
EventID	varchar	No	foreign key to tbl_Events (EventID)
WeatherID	int	No	automated identity
Weather	varchar	YES	descriptor from qry_tlu_Weather
Wind	tinyint	YES	wind speed - beaufort force from qry_tlu_WindScale
WindDirection	char	YES	cardinal direction
AirTemp_C	decimal	YES	air temperature in Celsius - degrees and tenths
AirTempTime	datetime	YES	time air temperature was measured
BarometricPressure_hg	decimal	YES	barometric pressure in inches of mercury (hg)
			tbl_Zooplankton - background information on how zooplankton survey was conducted
EventID	varchar	No	foreign key to tbl_Events (EventID)
LabID	varchar	YES	for tracking samples sent to lab - LakeCode + Date + MaxDepth
SampleID	int	No	Automated identity
CollectionTime	datetime	YES	time sample collected
UTME	int	YES	UTM Easting
UTMN	int	YES	UTM Northing
TowType	varchar	YES	tow type - vertical, horizontal from qry_tlu_LakesZooTowType
TowLength_m	decimal	YES	tow length in meters
NumberTowsCompleted	tinyint	YES	3 tows are required - built in table constraint
NetDiameter_cm	decimal	YES	net diameter in centimeters
NetLength_m	char	YES	net length in meters
SampleVolume	float	YES	sample volume
			tlu_AmphibSpecies - Lookup for NCCN Amphibian Species
Code	varchar	No	4-5 letter species code
Species	varchar	YES	species scientific name
Common_Name	varchar	YES	species common name
Class	varchar	YES	species class
Order	varchar	YES	species order
Family	varchar	YES	species family
			tlu_Substrate - Lookup for substrate types
SubstrateCode	char	No	single character code for substrate
SubstrateDescription	char	YES	description of substrate - bedrock cobble, boulder, grass, etc.
			tlu_Documents - Lookup for database/protocol documents
DocName	varchar	No	Document name/title
DocLink	varchar	YES	Hyperlink to document
Description	varchar	YES	Description/purpose of document
			tlu_EventsSubsurvey - Lookup for survey type
Subsurvey	varchar	No	name of survey type - Secchi, fish, zooplankton, etc.
			tlu_LifeStage - Lookup for amphibian species life stage
LifeStageCode	char	No	single character code for life stage
LifeStageDescrip	char	YES	Life-stage description - Adult, Egg, Larval, etc.

Acknowledgments

The authors would like to thank Andrea Woodward for her support during the development of this protocol; also Steve Fradkin, Roland Knapp, Kathleen Matthews, and Rolf Vinebrooke for reviewing and commenting on the previous draft of this document. A panel of scientists and resource managers (Mark Buktenica, Jordi Catalan, Stanley Dodson, David Donald, Steve Fradkin, Stanford Loeb, Paul Murtaugh, Tony Olsen, Barbara Samora, and Rolf Vinebrooke) also provided valuable comments and suggestions concerning the content of the protocol. We would also like to thank Reed Glesne, Bret Christoe, and Kim Anderson for their respective contributions, and Niels Leuthold for his contributions to the aquatic amphibian sections. Elisabeth Deimling, Gregg Lomnicky, Scott Girdner, Brendan Brokes, and William Warncke also deserve our thanks for spending countless hours in the backcountry areas of North Cascades National Park Service Complex and Mount Rainier National Park field testing and perfecting the sampling procedures described in this document. Funding for the development and completion of this manuscript was provided by the USGS Biological Resources Division in cooperation with the National Park Service. The National Park Service supported earlier field sampling and data collection that contributed to the development of the protocol.

References Cited

American Public Health Association, 1992, Standard methods for the examination of water and wastewater (18th ed.): Washington, D.C., American Public Health Association, variously paginated.

American Public Health Associaton, 1998, Standard methods for the examination of water and wastewater (20th ed.): Washington, D.C., American Public Health Association, variously paginated.

Bahls, P.F., 1991, Ecological implications of trout introductions to lakes of the Selway Bitterroot Wilderness, Idaho: Corvallis, Oregon State University, M.S. Thesis, 85 p.

Bahls, P., 1992, The status of fish populations and management of high mountain lakes in the western United States: Northwest Science, v. 66, p. 183-193.

Barbour, M.T., Gerritsen, J., Griffith, G.E., Frydenborg, R., McCarron, E., White, J.S., and Bastian, M.L., 1996, A framework for biological criteria for Florida streams using benthic macroinvertebrates: Journal of North American Benthological Society, v. 15, p. 185-211.

Barbour, M.T., Gerritsen, J., Snyder, B.D., and Stribling, J.B, 1997, Revision to Rapid Bioassessment Protocols for use in Streams and Rivers–Periphyton, Benthic Macroinvertebrates, and Fish: U.S. Environmental Protectin Agency, Washington, D.C., EPA 841-D-97-002, variously paginated.

Britton, L.J. and Greeson, P.E., eds., 1987, Methods for collection and analysis of aquatic biological and microbiological samples: U.S. Geological Survey Techniques of Water-Resource Investigations, book 5, Chap. A4, p. 131-144.

Brokes, B.J., 2000, Habitat segregation of two ambystomatids in mountain ponds, Mount Rainier National Park: Corvallis, Oregon State University, M.S. Thesis, 55 p.

Buktenica, M.W., and Larson, G.L., 1996, Ecology of kokanee salmon and rainbow trout in Crater Lake, Oregon: Lake and Reservoir Management, v. 12, p. 298-310.

Burnham, K.P., and Anderson, D.R., 2002, Model selection and inference—a practical information-theoretic approach: Springer-Verlag, New York, 488 p.

Cole, D.N., and Landres, P.B., 1996, Threats to wilderness ecosystems: impacts and research needs: Ecological Applications, v. 6, p. 168-184.

Corkran, C.C., and Thoms, C., 1996, Amphibians of Oregon, Washington and British Columbia: A field identification guide: Redmond, Wash., Lone Pine Publishing, 175 p.

Cuffney, T.F., Gurtz, M.E., and Meador, M.R., 1993, Methods for collecting benthic invertebrate samples as part of the National Water-Quality Assessment Program: U.S. Geological Survey Open-File Report 93-406, available online at http://water.usgs.gov/nawqa/protocols/OFR-93-406/inv1.html. (*Note*: *This report has been superseded by Moulton and others, 2002, available online at http://water.usgs.gov/nawqa/protocols/OFR02-150/index.html*)

Cummins, K.W., 1962, An evaluation of some techniques for the collection and analysis of benthic samples with special emphasis on lotic waters: American Midland Naturalist v. 67, p. 477-504.

Eilers, J.M., Brakke, D.F., Landers, D.H., and Overton, W.S., 1989, Chemistry of lakes in designated wilderness areas in the western United States: Environmental Monitoring and Assessment, v. 12, p. 3-21.

Franklin, J.F., 1987, Scientific use of wilderness, *in* Lucas, R.C., ed., Issues, state-of-knowledge, future directions: Proceedings of the National Wilderness Conference, Ogden, Utah: U.S. Forest Service General Technical Report INT-220, p. 42-46.

Franklin, J.F., and Dyrness, C.T., 1973, Natural vegetation of Oregon and Washington: Pacific Northwest Forest and Range Experiment Station, Portland, Ore., U.S. Department of Agriculture Forest Service General Technical Report PNW-8, 417 p.

Girdner, S.F., and Larson, G.L., 1995, Effects of hydrology on zooplankton communities in high-mountain ponds, Mount Rainier National Park, USA: Journal of Plankton Research v. 17, p. 1731-1755.

Gresswell, R.E., Liss, W.J., Lomnicky, G.A., Deimling, E.K., Hoffman, R.L., and Tyler, T., 1997, Using mark-recapture methods to estimate fish abundance in small mountain lakes: Northwest Science, v. 71, p. 39-44.

Hayslip, G.A., ed., 1993, Region 10 in-stream biological monitoring handbook for wadeable streams in the Pacific Northwest: U.S. Environmental Protection Agency, EPA/910/9-92-013, 75 p.

Hoffman, R.L., Liss, W.J., Larson, G.L., Deimling, E.K., and Lomnicky, G.A., 1996, Distribution of nearshore macroinvertebrates in lakes of the Northern Cascade Mountains, Washington, USA: Archiv fur Hydrobiologie v. 136, p. 363-389.

Hutchinson, G.E., 1957, A treatise on limnology, (vol. I), geography, physics, and chemistry: New York, John Wiley and Sons, Inc., 1015 p.

Karr, J.R., and Chu, E.W., 1997, Biological monitoring and assessment: using multimetric indexes effectively: Seattle, University of Washington, EPA 235-R97-001, 149 p.

Knapp, R.A., and Matthews, K.R., 2000, Non-native fish introductions and the decline of the mountain yellow-legged frog within protected areas: Conservation Biology, v. 14, p. 428-438.

Knapp, R.A., Matthews, K.R., and Sarnelle, O., 2001, Resistance and resilience of alpine lake fauna to fish introductions: Ecological Monographs, v. 71, p. 401-421.

Larson, G.L., and Hoffman, R.L. 2002, Abundances of northwestern salamander larvae in montane lakes with and without fish, Mount Rainier National Park, Washington: Northwest Science, v. 76, p. 35-40.

Larson, G.L., Wones, A., McIntire, C.D., and Samora, B., 1994, Integrating limnological characteristics of high mountain lakes into the landscape of a natural area: Environmental Management, v. 18, p. 871-888.

Larson, G.L., McIntire, C.D., Karnaugh-Thomas, E., and Hawkins-Hoffman, C., 1995, Limnology of isolated and connected high-mountain lakes in Olympic National Park, Washington State, USA: Archiv fur Hydrobiologie, v. 134, p. 75-92.

Larson, G.L., Lomnicky, G., Hoffman, R., Liss, W.J., and Deimling, E., 1999, Integrating physical and chemical characteristics of lakes into the glacially influenced landscape of the Northern Cascade Mountains, Washington State, USA: Environmental Management, v. 24, p. 219-228.

Leonard, W.P., Brown, H.A., Jones, L.L.C., McAllister, K.R., and Storm, R.M., 1993, Amphibians of Oregon and Washington: Seattle Audubon Society, Seattle, Wash.

Lewis, P.A., Klemm, D.J., and Thoeny, W.T., 2001, Perspectives on use of multimetric lake bioassessment integrity index using benthic macroinvertebrates: Northeastern Naturalist v. 8, p. 233-246.

Lind, O.T., 1979, Handbook of common methods in limnology (2nd ed.): The C.V. Mosby Co., London.

Liss, W.J., Larson, G.L., Deimling, E., Ganio, L., Gresswell, R., Hoffman, R., Kiss, M., Lomnicky, G., McIntire, C.D., Truitt, R., and Tyler, T., 1995, Ecological effects of stocked trout in naturally fishless high mountain lakes, North Cascades National Park Service Complex, Washington: National Park Service Technical Report NPS/PNROSU/NRTR-95-03, 285 p.

Loeb, S., 2002, Development of monitoring protocols for mountain lakes and ponds within the National Parks: Summary Report of Workshop, U.S. Geological Survey Forest and Rangeland Ecosystem Science Center, Corvallis, Oregon, August 7-9, 2002, 6 p.

Louter, D., 1998, Contested terrain: North Cascades National Park Service Complex, an administrative history: National Park Service. Seattle, Wash., 338 p.

MacKenzie, D.I., Nichols, J.D., Hines, J.E., Knutson, M.G., and Franklin, A.B., 2003, Estimating site occupancy, colonization, and local extinction when a species is detected imperfectly: Ecology, v. 84, p. 2200-2207.

MacKenzie, D.I., Nichols, J.D., Lachman, G.B., Droege, S., Royle, J.A., and Langtimm, C.A., 2002, Estimating site occupancy rates when detection probabilities are less than one: Ecology, v. 83, p. 2248-2255.

Merritt, R.W., Cummins, K.W., and Resh, V.H., 1984, Collecting, sampling, and rearing methods for aquatic insects, in Merritt, R.W., and Cummins, K.W., (eds.), An introduction to the aquatic insects of North America (2nd ed.): Dubuque, Iowa, Kendall/Hunt Publishing Co., p. 11-26.

Merritt, R.W., and Cummins, K.W., 1996, An introduction to the aquatic insects of North America (3rd ed.): Dubuque, Iowa, Kendall/Hunt Publishing Co., 862 p.

Morin, A., and Cattaneo, A., 1992, Factors affecting sampling variability of freshwater periphyton and the power of periphyton studies: Canadian Journal of Fisheries and Aquatic Sciences, v. 49, p. 1695-1703.

Moss, D., Furse, M.T., Wright, J.F., and Armitage, P.D., 1987, The prediction of macroinvertebrate fauna of unpolluted running-water sites in Great Britain using environmental data: Freshwater Biology, v. 17, p. 42-52.

Moulton, S.R., II, Kennen, J.G., Goldstein, R.M., and Hambrook, J.A., 2002, Revised protocols for sampling algal, invertebrate, and fish communities as part of the National Water-Quality Assessment Program: U.S. Geological Survey Open-File Report 02-150, available online at http://water.usgs.gov/nawqa/protocols/OFR02-150/index.html.

Nielsen, L.A., and Johnson, D.L., 1983, Fisheries techniques: Bethesda, Maryland, American Fisheries Society, 468 p.

Olson, D.H., Leonard, W.P., and Bury, R.B., (eds.), 1997, Sampling amphibians in lentic habitats: Northwest Fauna No. 4, Olympia, Wash., Society for Northwestern Vertebrate Biology, 134 p.

Parsons, D.J., and Graber, D.M., 1990, Horses, helicopters and hi-tech: managing science in wilderness *in* Proceedings of Preparing to manage wilderness in the 21st Century: U.S. Forest Service General Technical Report SE-66, p. 90-94.

Peine, J.D., 1990, The role of science in wilderness management *in* Proceedings of Preparing to manage wilderness in the 21st Century: U.S. Forest Service General Technical Report SE-66. p. 34-41.

Plafkin, J.L., Barbour, M.T., Porter, K.D., Gross, S.K., and Hughes, R.M., 1989, Rapid bioassessment protocols for use in streams and rivers: benthic macroinvertebrates and fish: U.S. Environmental Protection Agency, EPA/444/4-89-001.

Plotnikoff, R.W. and White, J.S., 1996, Taxonomic laboratory protocol for stream macroinvertebrates collected by the Washington State Department of Ecology: Olympia, Wash., Washington State Department of Ecology Publication No. 96-323, 32 p.

Porter, S.D., Cuffney, T.F., Gurtz, M.E., and Meador, M.R., 1993, Methods for collecting algal samples as part of the National Water-Quality Assessment Program: U.S. Geological Survey Open-File Report 93-409, 39 p.

Prepas, E.E., and Trew, D.O., 1983, Evaluation of the phosphorus-chlorophyll relationship for lakes off the Precambrian Shield in western Canada: Canadian Journal of Fisheries and Aquatic Sciences, v. 40, p. 27-35.

Schindler, D.W., 1987, Detecting ecosystem responses to anthropogenic stress: Canadian Journal of Fisheries and Aquatic Sciences, v. 44 (suppl. 1), p. 6-25

Stribling, J.B., Jessup, B.K., and Gerritsen, J., 2000, Development of biological and physical habitat criteria for Wyoming streams and their use in the TMDL process: Prepared for U.S. Environmental Protection Agency, Region 8, Denver, Colo. by Tetra Tech, Inc., Owings Mills, Maryland, 16 p.

Tyler, T., Liss, W.J., Ganio, L.M., Larson, G.L., Hoffman, R., Deimling, E., and Lomnicky, G., 1998, Interaction between introduced trout and larval salamanders (*Ambystoma macrodactylum*) in high-elevation lakes: Conservation Biology, v. 12, p. 94-115.

U.S. Environmental Protection Agency, 1998, Lake and reservoir bioassessment and biocriteria: Technical guidance document, U.S. Environmental Protection Agency, Office of Water, EPA 841-B-98-007, available online at http://www.epa.gov/owow/monitoring/tech/lakes.html.

U.S. Forest Service, 2003, Stream inventory handbook: Portland, Or. U.S. Department of Agriculture-U.S. Forest Service Region 6, 105 p.

Vinebrooke, R.D., and Leavitt, P.R., 1999, Phytobenthos and phytoplankton as bioindicators of climate change in mountain lakes and ponds: Journal of the North American Benthological Society, v. 18, p. 14-32.

Wetzel, R.G., and Likens, G.E., 2000, Limnological analyses (3rd ed.): Springer, New York, 429 p.

Wicklum, D.D., 1998, Effects of fish on lacustrine invertebrate community and seston dynamics. Missoula, University of Montana, PhD Dissertation, 85 p.

Appendix I. Sampling Equipment

General

Label Tape
Pencils and permanent markers (i.e., Sharpie)
Inflatable boat (2-3 person) and foot pump, or float tube and chest-waders
GPS unit
Nylon stuff sacks of variable sizes
Camera
Field data sheets and field book(s)
Rite-in-Rain® paper (8.5 x 11 in.)
Ziploc® bags (pint, quart, gallon)

SOP 1-4. Physical Characteristics, Bathymetry, Maximum Depth, Water-Level, Water Temperature, Water Clarity

Altimeter (if this function not available on GPS unit)
Planimeter or Electronic digitizer
Handheld depth sonar
Meter tape/meter-marked line or rope
Precision glass field thermometer
Thermistor/temperature recording device
Remote thermocouple
20 cm black and white Secchi disk
Parachute cord
Line-level
Measuring staff

SOP 5. Water Chemistry, Dissolved Organic Carbon, Dissolved Oxygen

Alpha® Water Bottle or Van Dorn-style water sampler
Meter-marked retrieval line
Messenger
250 mL-500 mL amber wide-mouth HDPE bottles
1 L HDPE bottle and aluminum extension pole for shallow sites
Forceps
Pre-washed 0.7 µm and 1.2 µm glass fiber filters (Whatman GF/F and GF/C)
500 mL Nalgene® polysulfone filtering apparatus (stand alone or bottletop) and hand pump or Nalgene® polysulfone syringe-type filter holder with 60 mL syringe
300 mL glass Wheaton bottles
60 mL amber glass bottlesLatex gloves and eye protection
$MnSO_4$ solution
Alkalai-iodide-azide reagent
Concentrated H_2SO_4
$Na_2S_2O_3$ (0.025 M)
Starch solution
Bi-iodate solution (0.0021 M)
10 mL – 100 mL pipettor with tips
1000 mL plastic buret with stop-cock
Nalgene® beaker (500 mL)

SOP 6. Chlorophyll-a Concentration and Periphyton

Alpha® Water Bottle or Van Dorn-type water sampler
Meter-marked retrieval line
Messenger
Any of the filtering apparatus set-ups describe above (i.e., SOP 5)
0.45 μm membrane filters
25 mL plastic scintillation vials
Forceps
1 L HDPE bottle and aluminum extension pole for shallow sites
Periphyton collection bottle (i.e., 250 mL wide-mouth HDPE bottle with bottom removed
Stiff-bristled brush
Pipettor and pipet tips
60 mL HDPE bottles
Plastic container 20 × 30 × 10 cm

SOP 7. Zooplankton (Rotifers and Crustaceans)

Marked retrieval line
Wisconsin-type zooplankton net (64 μm mesh, 20 cm opening, 1 m long)
Zooplankton cup that has sidewalls with openings covered with 64 μm mesh
60 mL HDPE bottles
95 percent ethanol (ETOH)
Small squirt bottle
Additional rope or parachute cord

SOP 8. Aquatic Macroinvertebrates

Meter tape/meter-marked line or rope
250 - 500 μm sieve
D-shaped sweep net
PVC-frame (1 m^2)
Forceps
Shallow-rectangular plastic container(s)
2 dram vials
WHIRL-PAK® (42 oz) or Ziploc® (1 gal) bags and garbage bags
70 percent ethanol (ETOH)
6-in. Ekman dredge (optional)
Plastic bucket or cubitainer (optional)

SOP 9. Aquatic Amphibian Sampling and Surveys

Personal Digital Assistant
Dry suit with neoprene hood, boots, and gloves (optional and intensive sites only)
Snorkel and mask (optional and intensive sites only)
Floatation device (optional)
Aquarium net
Surveyor's tape
Dive-light
Rule in cm and mm
Meter tape or meter-marked line or rope
Small plastic cubitainers for larval transport (if needed)
Amphibian identification key/field book

SOP 12. Fish Sampling and Surveys

Rod and reel
Terminal fishing tackle
Variable-mesh gill nets
Small scissors
5 gal collapsible plastic container
Shallow-rectangular plastic container
500 mL HDPE bottles
Measuring board (40 cm length, cm and mm)
Pesola™ spring scales (500 g and 1 kg)
Knife
Coin envelopes
WHIRL-PAK® (42 oz) or Ziploc® bags
Aluminum foil
95 percent ETOH
Plastic garbage bags
Pinesol®

Appendix II. Attributes to be Sampled During Each Sampling Period

July Sampling Period = JSP

August Sampling Period = ASP

Randomly Selected Level 2 Sites:

JSP: SOP 1; SOP 2 (bathymetry); SOP 7 and 8 (nearshore habitat/substrate assessment).
ASP: SOP 3 – SOP 8
JSP and **ASP:** SOP 2 (maximum depth and water-level); SOP 9; SOP 12

Note: *If there is to be only one sampling visit to the Level 2 sites, the sampling should occur in August and all SOPs should be completed at that time.*

Non-randomly Selected Level 1 Sites:

All attributes should be sampled during each sampling visit; except SOP 1 (attributes) and SOP 2 (bathymetry) should be completed during the first visit to the site.

This sampling schedule is provided only as an advisory schedule. At the discretion of the Project Supervisor this schedule can be amended. For instance, SOP 3 – SOP 8 attributes can be sampled during each sampling period if logistically and monetarily feasible.

Appendix III. Standard Measures and Units

SOP 1. Physical Characteristics of Monitoring Sites

Basin geologic composition: identify and record primary (i.e., dominant) rock types
Basin Watershed Area: hectares
Dominant Vegetation: types of vegetation to species when possible; however to lowest taxonomic level possible
Elevation: meters
Inlet Locations: UTM
Outlet Locations: UTM
Site Basin Aspect: compass direction
Site Basin Origin: type (see SOP 1 I)
Site Location: UTM
Site Perimeter: meters
Site Vegetation Zone: forest, subalpine, alpine
Surface Area: hectares

SOP 2. Bathymetry, Maximum Depth, Water-Level

Bathymetry Profile: depth in meters
Maximum Depth: meters
Water-Level: meters

SOP 3. Water Temperature

Temperature: Celsius (to level of precision of instrument used to make measurement)

SOP 4. Water Clarity

Secchi depth: meters

SOP 5. Water Chemistry, Dissolved Organic Carbon, Dissolved Oxygen

Alkalinity: mg/L
Ammonia: mg/L
Calcium: mg/L
Conductivity: µS/cm
Dissolved Organic Carbon: Total: mg/L
Dissolved Organic Carbon: Terrestrial Fraction: mg/L
Dissolved Organic Carbon: Attenuation: attenuation coefficient (K_d)
Dissolved Oxygen: mg/L
Magnesium: mg/L
Nitrate-N: mg/L
Orthophosphate: mg/L
pH: -log[H^+]; standard units 0-14

Potassium: mg/L
Sodium: mg/L
Sulfate: mg/L
Total Dissolved Solids: mg/L
Total Kjeldahl Nitrogen: mg/L
Total Phosphorus: mg/L

SOP 6. Chlorophyll-*a* Concentration and Periphyton

Chlorophyll-*a* : µg/L
Periphyton: Chlorophyll-*a* analysis: µg/L

SOP 7. Zooplankton (Rotifers and Crustaceans)

Taxa identification: to lowest taxonomic level possible
Biovolume: mg/m^3
Density: number/m^3

SOP 8. Aquatic Macroinvertebrates

Primary Substrates: type (see SOP 8 A 2)
Taxa identification: to lowest taxonomic level possible
Taxa: presence/absence; number/taxon/area sampled (i.e., m^2)

SOP 9. Aquatic Amphibian Sampling and Surveys

Taxa Identification: species
Taxa Life-Stage: embryo, larva or tadpole, metamorph, juvenile, adult
Taxa Size: snout-vent length in mm
Taxa Weight: grams
Population Elucidation: present/absent; PAO; number/transect (i.e., 25 m)

SOP 12. Fish Sampling and Surveys

Taxa Identification: species
Taxa Size: fork length in mm
Taxa Weight: grams
Population Elucidation: present/absent; estimated number/ha

Appendix IV. Sample Volumes

SOP 5. Water Chemistry, Dissolved Organic Carbon, Dissolved Oxygen

Water Chemistry Unfiltered Sample: 250–500 mL
Water Chemistry Filtered Sample: 500 mL
Dissolved Oxygen: approximately 1,000 mL unfiltered water
Dissolved Organic Carbon: 60 mL filtered water

SOP 6. Chlorophyll-*a* Concentration and Periphyton

Chlorophyl-*a* : minimum 500 mL filtered water
Periphyton sample: up to 60 mL sample and unfiltered water

SOP 7. Zooplankton (Rotifers and Crustaceans)

Sample: sample and unfiltered water to fill a 60 mL bottle.

Appendix V. Alternative Aquatic Amphibian Survey SOP for IS Sites

Version 1.00

Revision History Log:

Previous Version Number	Revision Date	Author	Changes Made	Reason for Change	New Version Number

This protocol is designed to be used as an alternative method for surveying amphibians at Intensively Sampled (IS) sites. The method can be used to generate observed abundance estimates for species and their reproductive effort. Visual Encounter Surveys (VES; Olson and others, 1997) and snorkel surveys (Tyler and others, 1998; Brokes, 2000) are effective for assessing the aquatic life stages of amphibians, whereas terrestrial life stages (e.g., juveniles and adults) are underrepresented in snorkel surveys. VES and snorkel surveys have typically been used to enumerate amphibians along the perimeter and in the nearshore areas of mountain ponds and lakes in the Pacific Northwest (Olson and others, 1997; Tyler and others, 1998; Brokes, 2000). Snorkel surveys are also useful for enumerating the presence/absence and observed abundances of amphibians in the deeper offshore area of a pond or lake, although these deep water surveys can be limited by the depth and offshore water clarity of the study site. Surveys can easily be conducted during the day, and can be performed at night with the aid of handheld dive-lights.

A. **Visual Encounter Survey:**

1. These procedures have been adapted from protocols developed by Mike Adams, USGS FRESC; also see Chapter 3 in Olson and others (1997).

2. The surveys are typically conducted by two people to determine the presence of amphibian species at a monitoring site.

3. The survey is conducted by both people walking the perimeter of the site and stopping every 10 m, or at each of 20 predetermined observation stations evenly spaced along the site perimeter, to visually search for amphibians. One person either wades, snorkels, or floats in the nearshore portion of the site, the other person walks on shore.

4. At each 10 m interval or at each observation station, a dipnet is used to capture visually detected amphibians to ensure definitive species identification. These individuals should be released at the point of capture.

5. Vegetated areas and other habitats likely to harbor amphibians should also be intermittently sampled using dipnets.

6. Once evidence for a species and particular life stage (i.e., embryos, tadpoles, larvae, metamorphs, adults) is detected and recorded at a site, searches for that species/life stage can be discontinued.

7. Dip-netting can be discontinued when all amphibian species potentially occurring at a site have been captured and identified.

8. If one or more species anticipated to occur at a site remain undetected following a complete circuit of the site, other habitats where the species are likely to occur (e.g., vegetated areas away from the shoreline) should be searched prior to declaring the species absent from the site.

9. All survey information should be recorded on the appropriate data sheet. See G below for information to be collected during each survey.

B. Snorkel Survey: Determination of Survey Transects (Tyler and others, 1998; Brokes, 2000):

1. During the first sampling visit to a monitoring site, four points are randomly selected along the perimeter of each site.

2. From each point, and in a clockwise direction, a 25 m transect along the site perimeter is measured using a meter tape or an incrementally marked length of rope. Each measured transect conforms to the contour of the shoreline.

3. Measured transects are marked at both ends with colored surveyor's tape.

4. General locations of survey transects are identified on a representational map of the study site and GPS coordinates for transect end points are recorded on the appropriate data sheet. This will facilitate relocation of transects on subsequent sampling visits.

5. For small sites (i.e., perimeter ≤ 100 m), the entire perimeter can be surveyed.

C. Snorkel Survey Gear:

1. A drysuit, neoprene hood, neoprene gloves, neoprene boots, mask, and snorkel, are essential for conducting surveys in often cold, mountain ponds and lakes.

2. Diving fins can be used but are not necessary and are often more troublesome than helpful in mountain pond and lake surveys.

3. A handheld dive-light can help illuminate areas in rock crevices and undercut shorelines that may harbor otherwise undetectable amphibians.

4. A segment of ruler can be affixed to or a ruler image can be inscribed on the dive-light to help the surveyor estimate total lengths of observed amphibians.

5. A small aquarium net and plastic container, usually carried by the data recorder assisting the surveyor, can be useful for capturing and holding amphibians for close inspection to facilitate species identification. These individuals should be released at the point of capture.

D. Conducting the Nearshore Snorkel Survey:

1. After survey transects have been determined, the snorkeler looks for amphibians in the open and by searching through substrate materials in the nearshore area of each transect.

2. The survey is conducted from shore to an approximate water depth of 1 m along each of the four survey transects.

3. The snorkeler reports amphibian species, number observed, estimated total lengths of individuals, and substrate type associated with observed amphibians to an onshore crew member recording information on a data sheet.

4. Snorkel surveys are not time constrained since the time to conduct a survey is dependent upon the complexity of the nearshore habitat.

5. A survey transect that is too shallow for effective snorkeling can be surveyed by walking/wading the transect in a zigzag pattern, to a depth of approximately 1 m (see Chapter 3 in Olson and others, 1997; Brokes, 2000).

6. All survey information should be recorded on the appropriate data sheet. See G below for information to be collected during each survey.

E. Conducting the Offshore Snorkel Survey:

1. Crew members should identify locations along the perimeter of a site where four transects perpendicular to the shoreline approximately 25 m in length can be accessed for deep water surveys.

2. These transects should begin approximately 3 to 5 m from the shoreline.

3. The snorkeler swims the transect counting the number of amphibians observed within the water column and on the bottom substrate of the deep water area being surveyed.

4. The crew member on shore will record the observations of the snorkeler and will notify the snorkeler when an estimated 25 m has been surveyed.

5. All survey information should be recorded on the appropriate data sheet. See G below for information to be collected during each survey.

F. Species Determination:

1. Amphibian species determination is performed in the field using distinguishing characteristics outlined in regional field guides (e.g., Leonard and others, 1993; Corkran and Thoms, 1996).

2. Species identification typically can be determined during the VES and snorkel survey. However, should identification prove difficult during a survey, individual amphibians can be captured for a closer inspection of key characteristics. These individuals should be released at the point of capture.

3. When amphibian identification proves too difficult to determine in the field, representative individuals can be transported from the field to a lab where they are reared to a size sufficient to confirm identification.

4. Live larvae are readily transported from the field in plastic containers provided water is kept cool and replenished when possible.

G. Information to be Collected and Recorded During Surveys:

1. Site Identifier.

2. Survey Date.

3. Type of Survey.

4. Weather Conditions.

5. Water and Air Temperature.

6. For each species detected:

    a. life stage (i.e., embryo, tadpole or larva, metamorph, adult).

    b. habitat (i.e., aquatic vegetation, woody debris, gravel, cobble, boulder, mud, silt, etc.).

    c. location (i.e., nearshore, shoreline, offshore, 10 m interval, or station; this information should be recorded on a representational map of the site and with GPS coordinates).

7. During snorkel surveys the number of individuals per species and their life stages should be recorded for each transect.

8. During VES, the number of individuals per species/ life stage observed can be estimated.

H. Guidelines for Use of Live Amphibians in Field Research:

1. Voucher specimens (generally a single individual of each species present) can be collected at each site if a voucher collection is identified as appropriate for the monitoring project. Individuals to be reared in the laboratory for species identification may also need to be collected.

2. Crew members should check to be certain that the project has applied for and received the appropriate state and federal collecting permits and that the project is in compliance with applicable Animal Care guidelines.

3. All crew members should have access to, read, and be familiar with the American Society of Ichthyologists and Herpetologists/Society for the Study of Amphibians and Reptiles publication "Guidelines for Use of Live Amphibians and Reptiles in Field Research". This publication is available only online at http://www.asih.org/pubs/herpcoll.html.

4. If it is not possible to identify larvae or embryos to species, a voucher specimen should be collected and stored in a glass vial containing 70 percent ethanol. If a voucher specimen is collected, the site identifier, collection date and time, specimen life-stage, habitat from which specimen was collected, and the name(s) of the individual(s) who collected the specimen should be recorded on the appropriate data sheet. The same information should be recorded, using a pencil on paper, and this tag should be placed into the vial with the specimen.

Amphibian Snorkel Survey Data Sheet

SITE: DATE:
PERSONNEL:

(Survey Type: NS=nearshore; OS=offshore. LHS=life history stage)

Survey Type	Transect No.	Species	LHS	Count	Habitat

Appendix VI. Field Data Sheets

The following data sheets are provided as outline summaries of the types of information that should be collected for each SOP and are not necessarily intended for use in the field. The information on the data sheets can be incorporated into revised data sheets that are more appropriate for field use or into appropriate PDA format.

LAKE MONITORING DATA SHEET (Page 1)   DATE:

I. GENERAL SITE INFORMATION AND PHYSICAL CHARACTERISTICS

1. **PERSONNEL:**

2. Site Identifier (e.g., name, park GIS code):

3. Watershed Identifier:

4. Time In:

5. Time Out (and date if different than Time In Date):

6. Site UTMe:

7. Site UTMn:

8. Weather Conditions:

9. Elevation (meters):

10. Surface Area (hectares):

11. Perimeter (meters):

12. Site Vegetation Zone (i.e., low forest, high forest, subalpine, alpine):

13. Dominant Basin Vegetation
    A. Trees:
    B. Shrubs:
    C. Herbs:
    D. Grasses:

14. Site Inlet(s) [include description(s) and UTM Coordinates]:

LAKE MONITORING DATA SHEET (Page 2)

SITE: DATE:

PERSONNEL:

I. GENERAL SITE INFORMATION AND PHYSICAL CHARACTERISTICS—Continued

15. Site Outlet(s) [include description and UTM Coordinates]:

16. Site Basin Aspect:

17. Site Basin Watershed Area:

18. Maximum Depth (meters):

    A.   Depth:

    B.   UTMe:

    C.   UTMn:

19. Bathymetric Map and/or Measurements Completed (Y/N):

20. Air Temperature (Celsius) and Time:

21. Water Temperature (Celsius) Profile (completed at Maximum Depth UTM Coordinates):

Time: Temp: Depth:

    A.   below-surface:

    B.   mid-column:

    C.   above-bottom:

22. Secchi Depth (completed at Maximum Depth UTM Coordinates):

Weather and water surface conditions:

Time:

1a) descend depth:	2a) descend depth:	3a) descend depth:
1b) ascend depth:	2b) ascend depth:	3b) ascend depth:
1c) mean depth:	2c) mean depth:	3c) mean depth:

LAKE MONITORING DATA SHEET (Page 3)

SITE: DATE:
PERSONNEL:

I. GENERAL SITE INFORMATION AND PHYSICAL CHARACTERISTICS—Continued

23. Basin Mineral Composition (samples collected yes?; no?):

24. Site Basin Origin (check one):

    A.   Bench:

    B.   Cirque:

    C.   Tarn:

    D.   Ice Scour:

    E.   Kettle:

    F.   Moraine (I.e., lateral, terminal, both):

    G.   Slump:

    H.   Trough:

    I.   Fault-influenced:

25. Visual Record (i.e., 35 mm, digital):

    A.   Film Roll or Disk Number:

    B.   Exposure Number:

LAKE MONITORING DATA SHEET (Page 4)

SITE: DATE:

PERSONNEL:

II. WATER QUALITY (UNFILTERED SAMPLE)

Collected at Maximum Depth UTM Coordinates

1. Sample Number:

2. Collection Time:

3. Sample Depth (meters):

4. Sample Volume (milliliters):

III. WATER CHEMISTRY (FILTERED SAMPLE)

Filter size = 1.2 μm GF/C;   Collected at Maximum Depth UTM Coordinates

1. Sample Number:

2. Collection Time:

3. Sample Depth (meters):

4. Sample Volume (milliliters):

IV. DISSOLVED ORGANIC CARBON (FILTERED SAMPLE)

Filter size = 0.7 μm GF/F;   Collected at Maximum Depth UTM Coordinates

1. Sample Number:

2. Collection Time:

3. Sample Depth (meters):

4. Sample Volume (milliliters):

V. DISSOLVED OXYGEN (DO)

1. Beginning volume $Na_2S_2O_3$ (milliliters):

2. Ending volume $Na_2S_2O_3$ (milliliters):

3. Total volume $Na_2S_2O_3$ titrated (milliliters):

4. mgDO/L:

LAKE MONITORING DATA SHEET (Page 5)

SITE: DATE:
PERSONNEL:

VI. BIOLOGICAL SAMPLES

A. *Chlorophyll-a Water Sample (filtered)*

Filter size = 0.45 μm membrane;   Collected at Maximum Depth UTM Coordinates

   1.   Sample Number:

   2.   Collection Time:

   3.   Sample Depth (meters):

   4.   Sample Volume Filtered (mL):

B. *Periphyton (epilithic)*

1a) Sample Number:	5a) Sample Number:
1b) Collection Time:	5b) Collection Time:
1c) Sample UTMe:	5c) Sample UTMe:
1d) Sample UTMn:	5d) Sample UTMn:
1e) Sample Substrate:	5e) Sample Substrate:
2a) Sample Number:	6a) Sample Number:
2b) Collection Time:	6b) Collection Time:
2c) Sample UTMe:	6c) Sample UTMe:
2d) Sample UTMn:	6d) Sample UTMn:
2e) Sample Substrate:	6e) Sample Substrate:
3a) Sample Number:	7a) Sample Number:
3b) Collection Time:	7b) Collection Time:
3c) Sample UTMe:	7c) Sample UTMe:
3d) Sample UTMn:	7d) Sample UTMn:
3e) Sample Substrate:	7e) Sample Substrate:
4a) Sample Number:	8a) Sample Number:
4b) Collection Time:	8b) Collection Time:
4c) Sample UTMe:	8c) Sample UTMe:
4d) Sample UTMn:	8d) Sample UTMn:
4e) Sample Substrate:	8e) Sample Substrate:

LAKE MONITORING DATA SHEET (Page 6)

SITE: DATE:

PERSONNEL:

VI. BIOLOGICAL SAMPLES—Continued

C. Zooplankton Samples

1a) Sample Number:
1b) Collection Time:
1c) Sample UTMe:
1d) Sample UTMn:
1e) Tow Type (vert; horiz):
1f) Tow Length (meters):

2a) Sample Number:
2b) Collection Time:
2c) Sample UTMe:
2d) Sample UTMn:
2e) Tow Type (vert; horiz):
2f) Tow Length (meters):

3a) Sample Number:
3b) Collection Time:
3c) Sample UTMe:
3d) Sample UTMn:
3e) Tow Type (vert; horiz):
3f) Tow Length (meters):

D. *Aquatic Macroinvertebrate Samples*

(ns = nearshore; os = offshore; pr = preserved; pk = picked)

1a) Sample Number:
1b) Sample Type (ns; os):
1c) Collection Time:
1d) Sample UTMe:
1e) Sample UTMn:
1f) Habitat Primary Substrate:
1g) Habitat Secondary Substrate:
1h) Process method (pr; pk):
1i) Subsampled?:
1j) Estimate subsample amount:

2a) Sample Number:
2b) Sample Type (ns; os):
2c) Collection Time:
2d) Sample UTMe:
2e) Sample UTMn:
2f) Habitat Primary Substrate:
2g) Habitat Secondary Substrate:
2h) Process method (pr; pk):
2i) Subsampled?:
2j) Estimate subsample amount:

3a) Sample Number:
3b) Sample Type (ns; os):
3c) Collection Time:
3d) Sample UTMe:
3e) Sample UTMn:
3f) Habitat Primary Substrate:
3g) Habitat Secondary Substrate:
3h) Process method (pr; pk):
3i) Subsampled?:
3j) Estimate subsample amount:

4a) Sample Number:
4b) Sample Type (ns; os):
4c) Collection Time:
4d) Sample UTMe:
4e) Sample UTMn:
4f) Habitat Primary Substrate:
4g) Habitat Secondary Substrate:
4h) Process method (pr; pk):
4i) Subsampled?:
4j) Estimate subsample amount:

LAKE MONITORING DATA SHEET (Page 7)

SITE: DATE:
PERSONNEL:

VI. BIOLOGICAL SAMPLES—Continued

D. *Aquatic Macroinvertebrate Samples (Continued)*

(ns = nearshore; os = offshore; pr = preserved; pk = picked)

5a) Sample Number:

5b) Sample Type (ns; os):

5c) Collection Time:

5d) Sample UTMe:

5e) Sample UTMn:

5f) Habitat Primary Substrate:

5g) Habitat Secondary Substrate:

5h) Process method (pr; pk):

5i) Subsampled?:

5j) Estimate subsample amount:

6a) Sample Number:

6b) Sample Type (ns; os):

6c) Collection Time:

6d) Sample UTMe:

6e) Sample UTMn:

6f) Habitat Primary Substrate:

6g) Habitat Secondary Substrate:

6h) Process method (pr; pk):

6i) Subsampled?:

6j) Estimate subsample amount:

7a) Sample Number:

7b) Sample Type (ns; os):

7c) Collection Time:

7d) Sample UTMe:

7e) Sample UTMn:

7f) Habitat Primary Substrate:

7g) Habitat Secondary Substrate:

7h) Process method (pr, pk):

7i) Subsampled?:

7j) Estimate subsample amount:

8a) Sample Number:

8b) Sample Type (ns; os):

8c) Collection Time:

8d) Sample UTMe:

8e) Sample UTMn:

8f) Habitat Primary Substrate:

8g) Habitat Secondary Substrate:

8h) Process method (pr; pk):

8i) Subsampled?:

8j) Estimate subsample amount:

LAKE MONITORING DATA SHEET (Page 8)

SITE: DATE:

VII A. FISH SAMPLING: MARK-RECAPTURE

Activity	M1	M2	M3	R1	R2	R3	C

M1 = Number fish marked during initial marking visit; usually adipose fin-clip
M2 = Number fish marked during second marking visit; usually upper caudal fin clip
M3 = Number fish marked during third marking visit; usually lower caudal fin clip
R1 = Number of recaptures of M1 at each net check
R2 = Number of recaptures of M2 at each net check
R3 = Number of recaptures of M3 at each net check
C = Total number of captutres (marked + unmarked) at each net check
M1 rel, M2 rel, M3 rel = Number of M1, M2, M3 fish released alive at each net check
UM rel = Number of unmarked fish released alive at each net check

LAKE MONITORING DATA SHEET (Page 9)

SITE: DATE:
PERSONNEL:

VII B. FISH SAMPLING (Non M-R)   FISH PRESENT (?):

Sample No.	Species	Sex: Male	Female	Unknown	Total Length (mm)	Weight (g)

LAKE MONITORING DATA SHEET (Page 10)

SITE:

VIII. DATA FIELD SHEETS COMPLETION VERIFICATION:

A. Data Field Sheets Checked By:

B. Date:

C. Time:

Appendix VII. Amphibian Survey Data Sheets for Collection of PAO Data With Variable Definitions

Visual Encounter Survey for Amphibians   **Page 1**

Field QA _____   Entered By _____

SITE LOCATION:

Site ID		Drain-age			Quad name	
State		Co.		Owner-ship		Datum	
Lat			Lon			PDOP	
Elev.		m / ft	GPS File				
Loca-tion							

SITE VISIT:

Date		Visit #		Crew	
Start Time		End Time		Search Minutes (minutes times # of surveyors)	

Weather (check one)	Wind (check one)	Air temp		F C
❏ Clear or few clouds	❏ Calm	Water temp		F C
❏ Cloudy or overcast	❏ Light breeze			
❏ Showers or light rain	❏ Gusts	Water present?	Fish present?	
❏ Heavy rain	❏ Windy	Y N	Y N	
❏ Sleet or snow				

POND:

Length (m)		Width (m)		Max depth (m)	<1 1-2 >2	% shallow (<0.5m)	
Color (check one)	❏ Clear ❏ Stained	Trans-parency (check one)	❏ Clear ❏ Opaque	% emergent veg cover		% of site edge searched	

Wetland Type (check one)	Site origin (check one)	Beaver sign (check all that apply)	Permanence (check one)
❏ Lake/pond		❏ Saw beaver	
❏ Meadow/Wetland	❏ Naturally occurring	❏ Heard beaver	❏ Permanent
❏ Ditch		❏ Current dam	
❏ Beaver pond	❏ Manmade/artificial	❏ Old dam	❏ Semi-permanent
❏ River/Stream		❏ Cuttings	
❏ Spring/Seep	❏ Human altered	❏ Lodge	❏ Temporary
❏ Oxbow/Backwater		❏ None	

Visual Encounter Survey for Amphibians   Page 2

Map	N ↑

Mark Locations: S = water samples, T = temperature, G = GPS location, P = photograph
Sketch emergent vegetation using dotted lines.

Dominant Substrate & Percent Veg Cover (up to 20 transects per site)					
% Cover					
Substrate (circle one per transect)	CI Si M Sa G P Co B L W	CI Si M Sa G P Co B L W	CI Si M Sa G P Co B L W	CI Si M Sa G P Co B L W	CI Si M Sa G P Co B L W
% Cover					
Substrate (circle one per transect)	CI Si M Sa G P Co B L W	CI Si M Sa G P Co B L W	CI Si M Sa G P Co B L W	CI Si M Sa G P Co B L W	CI Si M Sa G P Co B L W
% Cover					
Substrate (circle one per transect)	CI Si M Sa G P Co B L W	CI Si M Sa G P Co B L W	CI Si M Sa G P Co B L W	CI Si M Sa G P Co B L W	CI Si M Sa G P Co B L W
% Cover					
Substrate (circle one per transect)	CI Si M Sa G P Co B L W	CI Si M Sa G P Co B L W	CI Si M Sa G P Co B L W	CI Si M Sa G P Co B L W	CI Si M Sa G P Co B L W

Substrates: Clay, Silt, Mud/organic muck, Sand, Gravel, Pebble, Cobble, Boulder/bedrock,
Leaf litter, downed Wood

Visual Encounter Survey for Amphibians   Page 3

Site ID		Drainage		Date	

SPECIES						
Species code	Life stage	SVL/TL (mm)	Count	Count method (circle one)	Calling (circle one)	Detection method (circle one)
				A E	Y N	H V A
Notes:						
				A E	Y N	H V A
Notes:						
				A E	Y N	H V A
Notes:						
				A E	Y N	H V A
Notes:						
				A E	Y N	H V A
Notes:						
				A E	Y N	H V A
Notes:						
				A E	Y N	H V A
Notes:						
				A E	Y N	H V A
Notes:						
				A E	Y N	H V A
Notes:						
				A E	Y N	H V A
Notes:						
				A E	Y N	H V A
Notes:						
				A E	Y N	H V A
Notes:						
				A E	Y N	H V A

Life Stage: Egg, EM Egg Mass, Larval, Metamorph, Juvenile, Paedomorph, Adult. Count method: number recorded is Actual or Estimated. Detection Method Hand Collected (catch and see), Visual (see from a distance only), Aural (heard calling only).

Visual Encounter Survey for Amphibians   **Page 4**

NOTES:

Site Location Definitions for Data Sheet Page 1

Field	Definition
Site ID	Record the site name. The site name should be unique to the study area.
Drainage	River drainage in which the site is located.
Quad name	Name of the USGS 7.5' topographic quadrangle map on which one would find the survey site.
State	Enter the two-letter abbreviation of the state in which the survey site is located.
County	Enter name of the county in which the site is located.
Ownership	Indicate who owns or manages the land on which the site is located.
Datum	Enter the datum used to derive latitude/longitude coordinates. The datum can be found on the GPS unit, and should always be WGS84.
Latitude/Longitude	This is a pair of numbers that are x and y coordinates. Location of each site will be recorded in decimal degrees using a hand held GPS unit in the field. For lakes and ponds, enter the coordinates for the edge of the water if you are using a GPS or from the center of the water body if you are obtaining the coordinates from a USGS 7.5' topographic map. If using a GPS unit, mark on the site map where the reading was taken.
PDOP	Record the error estimate supplied by the GPS unit.
Elevation	Record the elevation (in meters) derived from a 7.5' topographic map or GIS coverage layer. Elevation should not be taken from the GPS unit because some units are not sufficiently accurate.
GPS file	If the waypoint for a site was saved on the GPS unit, record the name of the file.
Location	This is a description that would allow someone not familiar with the area to find the pond again. For example: "200 m NE of Lake Constance." Also, it is a good place to put the name of a lake if it has an official name (e.g., Hidden Lake). Use mapped landmarks that are not likely to change. Some localities are difficult to describe and the latitude/longitude coordinates will be the most descriptive, but a narrative description is important to confirm the general location.

Site Visit Definitions for Data Sheet Page 1

Field	Definition
Date	Write the date as day, 3-letter abbreviation for month, and 4-digit year (e.g. 11 Aug 2003). The three letter abbreviation for the month is less ambiguous and more readily recognized than 8-11-03.
Visit number	Several sites will be visited more than one time in a field season. Record the visit number of the survey being conducted for the current field season.
Crew	Enter the names of the crewmembers present, beginning with the person filling out the datasheet.
Start time	Record the time that your survey began, not time you arrived at the site.
End time	Record the time when the survey ended, not time you finished taking notes and gathering up equipment.
Search minutes	This is the total time (in minutes) spent searching for amphibians multiplied by the number of active observers. Do not include time spent processing specimens, recording notes, or taking photographs. If two or more people spend differing amounts of time searching, use the average time.
Weather	Check <u>only one</u> of the five weather choices: *Clear or few clouds* – less than 50% cloud cover, *Cloudy or overcast* – greater than 50% cloud cover, *Showers or light rain* – light or sporadic rainfall, *Heavy rain* – consistent rainfall, or *Sleet or snow* – consistent sleet or snow fall.
Wind	Check <u>only one</u> of the four wind choices: *Calm* – little or no wind, smoke rises vertically; *Light breeze* – wind felt on face, leaves rustle, wind relatively consistent; *Gusts* – wind varies with periods of little wind and periods of higher wind, higher wind is strong enough to keep leaves and twigs in constant motion; *Windy* – relatively consistent wind that is at least strong enough to move small branches and cause dust to rise.
Air temperature	Record the air temperature (in ° C) in the shade at approximately 1 meter off the ground.
Water temperature	Record the water temperature (in ° C) near the surface either in the vicinity of any observed amphibians, or approximately 0.5 meters out from the edge of the water. Since temperature can be quite variable in standing water, try to select a location that is representative of the site.
Water present	Circle *Y* if there is water present at the survey site and *N* if the survey site is dry.
Fish present	Circle *Y* if there are fish detected at the survey site and *N* if fish are not detected at the survey site.

Pond Definitions (all measurements in meters) for Data Sheet Page 1

Field	Definition
Length	Estimate the length of the site in meters.
Width	Estimate of the width of a site in meters. This estimate should be based on a line that is perpendicular to the length.
Max Depth	If the maximum depth is less than 1 m circle *<1*. If the maximum depth is between 1 and 2 m, circle *1-2*. If the maximum depth is greater than 2 m, circle *>2*.
% shallow	Record the percentage of the wetted site area that is less than 0.5 m deep.
Color	Check <u>only one</u> of the two choices for water color: *Clear* or *Stained*. Stained water looks like weak coffee or tea.
Transparency	Check <u>only one</u> of the two choices for the water transparency: *Clear* or *Opaque*.
% emergent veg. cover	Record the percent emergent/floating vegetation for the portion of the site visible to you from the shore. This should only include the vegetation that is actually rooted or floating in the water and touches the surface.
% of site edge searched	Record an estimate of how much of the edge of the site was searched. For small sites this will often be 100%.
Wetland type	Check <u>only one</u> of the seven choices that best describes the type of wetland being surveyed: *Pond/lake, River/stream, Meadow/wetland, Spring/seep, Ditch, Oxbow/backwater,* or *Beaver pond.*
Site origin	Check <u>only one</u> of the choices of site origin: *Natural, Man-made* or *Altered.*
Beaver sign	Check <u>one or more</u> of the seven options if any of the listed beaver signs were present: *Saw Beaver, Current Dam, Heard Beaver, Old Dam, Lodge, Cuttings,* and *None.*
Permanence	Check <u>only one</u> of the three options that best describes site permanence: *Permanent* – water never dries, *Semi-permanent* – water dries in some years but not annually, or *Temporary* – water dries annually.

Map and Substations Definitions for Data Sheet Page 2

Field	Definition
Map	Draw a sketch of the site. Use a compass to make sure sketch is oriented to North arrow on map. Indicate on map where water samples, GPS readings and temperature measurements took place (use *S* to indicate water samples, *T* to indicate the location of temperature measurements, *G* to indicate position of GPS measurement, and *P* to indicate where photograph was taken). Indicate habitat details such as large patches of emergent vegetation, logs or other important habitat features occur.
% cover	Estimate the percent of the surface area with floating or emergent aquatic vegetation for a 2 × 2 meter box that extends from the shore out into the water. This should be estimated at each station established for the amphibian survey. If the site is small, make one estimate for the entire site.
Substrate	Circle <u>only one</u> of the 10 choices for the dominant substrate for the 2 × 2 meter box at each station: *C* – <u>C</u>lay, *Si* – <u>Si</u>lt, *M* – <u>M</u>ud/organic muck, *Sa* – <u>Sa</u>nd, *G* – <u>G</u>ravel, *P* – <u>P</u>ebble, *Co* – <u>Co</u>bble, *B* –<u>B</u>oulder/bedrock, *L* – <u>L</u>eaf litter, *W* – downed <u>W</u>ood. This should be estimated at each station established for the amphibian survey. If the site is small make one estimate for the entire site.

Species Definitions for Data Sheet Page 3

Field	Definition
Site ID	A unique identifier for the site being surveyed. This should be identical to the Site ID recorded on page 1.
Drainage	River drainage in which the site is located. This should be identical to the drainage recorded on page 1.
Date	Write the date as day, 3-letter abbreviation for month, and 4-digit year (e.g. 11 Aug 2003). The three letter abbreviation for the month is less ambiguous and more readily recognized than 8-11-03. This should be identical to the date recorded on page 1.
Species code	Record the code for the species of amphibian. See SOP 9 for species names and codes. A separate line should be used for each species and life stage of a species that is observed. If size classes are apparent, a different line should be used for each size (e.g., young-of-the-year vs. second-year larvae).
Life Stage	For amphibian species record the life stage of the group being counted. The possible life stages are E – Egg, EM – Egg Mass, L – Larval, M – Metamorph, J – Juvenile, P – Paedomorph, or A – Adult.
SVL/TL (mm)	Record the snout-to-vent and total length (in millimeters) for captured individuals from the species/life stage group. Record lengths for individual animals by separating SVL and TL with a slash (e.g., 53/100). Measure 10 individuals if possible.
Count	Record the number of individuals of a species and life stage that are seen. If tally marks are used to keep a running count, add them up at the end of the survey. Record this number and circle it.
Count method	Circle A or E to indicate if the count is actual or estimated.
Calling	Circle Y if the amphibian was heard calling and N if the amphibian was not calling.
Detection method	Circle one of the three choices to indicate how individuals were encountered: H – Hand Collected, V – Visual (seen from a distance only), A – Aural (only heard calling).
Notes	Use the notes section to record anything unusual.

Definitions for Data Sheet Page 4

Field	Definition
Notes	Record notes on anything of interest. It is especially important to record any deviations for the normal sampling protocol. If a datasheet field that is normally filled out is left blank, please indicate why the data could not be collected (e.g., "the thermometer was broken so temperature was not recorded"). Use this space to record the presence of any unusual or interesting species or site conditions.

www.ingramcontent.com/pod-product-compliance
Lightning Source LLC
Chambersburg PA
CBHW081549170526
45166CB00009B/2634